Geometric and Engineering Drawing

This introduction to descriptive geometry and contemporary drafting guides the student through the essential principles to create engineering drawings that comply with international standards of technical product specification. This heavily updated new edition now applies to CAD as well as conventional drawing. Extensive new coverage is given of:

- International drafting conventions
- Methods of spatial visualisation such as multi-view projection
- Types of views
- Dimensioning
- Dimensional and geometric tolerancing
- Representation of workpiece and machine elements
- Assembly drawings

Comprehensible illustrations and clear explanations help the reader master drafting and layout concepts for creating professional engineering drawings. The book provides a large number of exercises for each main topic. This edition covers updated material and reflects the latest ISO standards.

It is ideal for undergraduates in engineering or product design, students of vocational courses in engineering communication and technology students covering the transition of product specification from design to production.

Ken Morling trained as a mechanical engineer at Vickers Armstrong in 1956. He helped develop many aircraft including the TSR2 and Concorde. In 1963 he started teaching technical drawing to GCE O and A level students and wrote this book first published in 1969. In 1965 he became a graduate of the Institution of Mechanical Engineers and a Master of Philosophy in 1980 for development in education. During the life of his book there have been three editions and two translations into Spanish and Portuguese.

Stéphane Danjou is Professor of Mechanical Engineering and Plant Design at Rhine-Waal University of Applied Sciences in Germany, teaching Engineering Drawing, 3D CAD and Design. He started as a draftsman in a global acting company, studied mechanical engineering and worked as Technical Director in the packaging industry. With more than 25 years of experience in engineering drawing, he specialises in product development, particularly within the context of modern approaches in engineering design.

Geometric and Engineering Drawing

Fourth Edition

Ken Morling
Stéphane Danjou

Routledge
Taylor & Francis Group

LONDON AND NEW YORK

Cover credit: iStock/matejmo

Fourth edition published 2022
by Routledge
4 Park Square, Milton Park, Abingdon, Oxon, OX14 4RN

and by Routledge
605 Third Avenue, New York, NY 10158

Routledge is an imprint of the Taylor & Francis Group, an informa business

First edition published by Edward Arnold 1969
Third edition published by Routledge 2010

British Library Cataloguing-in-Publication Data
A catalogue record for this book is available from the British Library

Library of Congress Cataloging-in-Publication Data
A catalog record has been requested for this book

ISBN: 978-0-367-43127-3 (hbk)
ISBN: 978-0-367-43123-5 (pbk)
ISBN: 978-1-003-00138-6 (ebk)

DOI: 10.1201/9781003001386

Typeset in Sabon
by Apex CoVantage, LLC

Contents

Preface

Engineering drawings have been used to convey clear information about an object or system for many decades. They are graphic representations of products and depict how an object is designed, how it functions and how it is supposed to be manufactured and assembled. With the advent of comprehensive computer-aided design (CAD) programs, the way engineering drawings are created significantly changed. The introduction of three-dimensional (3D) CAD systems especially has revolutionised the approach for conveying the specifications of a technical product. Although the tools for creating engineering drawings have changed dramatically with time, the underlying principles are still the same. The fundamentals of descriptive geometry are still necessary for many applications of geometric drawing, and the same conventions and best practices still apply to the drafting process. Despite the rapid pace of computational improvements and the resulting possibilities of supporting and rationalising the drafting process, the basic concepts of geometric and engineering drawing have lost nothing of their relevance. Nowadays, geometric drawing is an essential skill for graphically solving geometric problems as part of a concept phase in a design process and for understanding the theory behind CAD systems. Although the current trend is to communicate product data exclusively with means of 3D CAD models for paperless production, we are far away from sending engineering drawings into retirement. Engineering drawings, no matter if resulting from manual drafting or as end products of a CAD modelling process, are still an efficient form of communication, required in all engineering disciplines.

This fourth edition has been entirely revised and heavily updated. While the fundamentals of geometric drawing have been taken from earlier editions, some completely new chapters are included that emphasise each part of the engineering drawing.

The new edition starts with an introduction to engineering communication and some basics of engineering drawing such as drafting equipment, internationally agreed conventions and principal techniques of technical sketching. The chapter on geometric constructions is almost unchanged from earlier editions. It shows specific drawing solutions to many geometric problems. Before introducing readers to the different types of views in engineering drawing, methods of spatial visualisation and basic concepts of descriptive geometry are presented. New chapters thoroughly address the important topics of dimensioning and tolerancing. Further, we introduce common representations of workpiece elements and frequently occurring machine elements as well as assembly drawings and their properties.

All chapters related to engineering drawing have been updated to the latest recommendations from the International Organization for Standardization (ISO), that is, best practices from the international standards developing organisation.

However carefully one checks a manuscript, errors creep in. We shall be very grateful if any readers who find errors let us know through the publishers.

S. Danjou & K. Morling
November 2021

Acknowledgements

To start with, I wish to express my sincere appreciation for the excellent work Ken Morling has provided with all previous editions of this book. I am deeply grateful for the opportunity to carry on with the success story he started more than 50 years ago.

My special thanks to the following examination boards for giving their permission to use questions from past papers. I am particularly grateful to them for allowing us to change many of the questions from imperial to metric units.

Certificate of Secondary Education

Associated Lancashire Schools Examining Board
East Anglian Regional Examinations Board
Metropolitan Regional Examinations Board
Middlesex Regional Examining Board
North Western Secondary School Examinations Board
South-East Regional Examinations Board
Southern Regional Examinations Board
West Midlands Examinations Board

General Certificate of Education

Associated Examining Board
Local Examinations Syndicate, University of Cambridge
Joint Matriculation Board
University of London School Examinations
Oxford Delegacy of Local Examinations
Oxford and Cambridge Schools Examination Board
Southern Universities' Joint Board

I especially thank the West Midlands Examinations Board, the Associated Lancashire Schools Examining Board and the Southern Universities' Joint Board for allowing us to draw solutions to questions set by them.

The extracts from BS 308, Engineering Drawing Practice and BS 3692, ISO Metric Precision Hexagon Bolts, Screws and Nuts are taken from a number of recent British Standards Institution Publications who have given their permission for the reproductions. Copies of the complete standards are available from BSI, 2, Park Street, London W1A 2BS.

My sincere thanks to the publisher, Taylor & Francis, and its editorial staff for making this fourth edition possible, and for showing so much sympathy during these difficult times which affected so many of us.

I would like to express heartfelt gratitude to my beloved wife, Marie-Kristin, and my children Lisanne and Eric for all their support, patience and encouragement. Finally, this edition became reality because they tolerated my incessant disappearances into my home office.

S. Danjou

The fourth edition has been compiled by Stéphane Danjou. I am very grateful that he has offered me many opportunities to make contributions where he thought they were appropriate. My thanks to S. Pagett for checking the manuscript of the original publication.

K. Morling

Chapter 1

Introduction to Engineering Communication

1.1 PRODUCT DEVELOPMENT AND THE ENGINEERING DESIGN PROCESS

There are many different definitions for the engineering design process. However, they all share common attributes. In the broad view, engineering design, sometimes also referred to as technical design, is a systematic process where basic sciences such as mathematics and physics as well as engineering sciences are applied to solve a given problem. So it is about devising a component or a system while taking predetermined requirements into account to meet the desired needs.

Usually, engineering design follows a well-defined sequence of process steps. From a macroscopic perspective, the engineering design process can be broken down into four stages (see Figure 1.1).

At the very beginning, a design engineer is confronted with a problem statement. Note that a technical problem is given when the solution is not available with the help of already known means. As a result, it is the task of an engineer to establish a clear task description by clarifying the frame conditions and specifying the requirements. The latter can result from customer requirements such as performance, ergonomic or aesthetic requirements, from internal requirements such as considerations with respect to manufacturing or costs, or from external requirements like social and regulatory issues. The complete set of requirements represents constraints the developed technical solution will need to satisfy.

In a second step, the challenge is to develop one or more concepts that have the potential to solve the given problem. This is usually done by establishing a desired function structure and searching for working principles that fulfil the subfunctions. When combining the found working principles into so-called working structures while taking technical and economic constraints into consideration, we will get a set of principle solution variants.

Once principle solutions have been found that meet given evaluation criteria, the design process steps forward into an embodiment design phase where key modules are defined. At this stage, the size of product features or assembly components as well as their arrangement are determined. The outcome of the

DOI: 10.1201/9781003001386-1

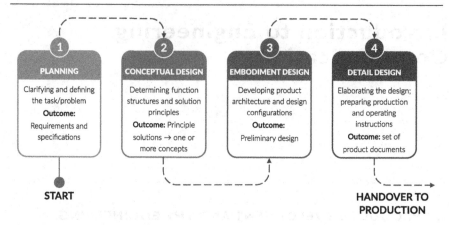

Figure 1.1 The engineering design process.

embodiment design are detailed layouts with respect to main function carriers and auxiliary function carriers. In addition to the basic rules of embodiment design 'clarity', 'simplicity', 'sustainability' and 'safety', numerous so-called principles of embodiment design and guidelines, depending on the selected manufacturing process, have to be taken into account. Further, the design is checked against errors, any kind of disturbing factors and possible risks.

A subsequent detail design process step, where all design details are finalised and where compliance with initial requirements are checked with the help of prototypes and tests, results into a set of product documents, partly needed for production and operation. This can include:

- Manufacturing drawings
- Assembly drawings
- Bill of materials (BOM), also known as parts lists
- Production/assembly/transportation instructions
- User manual
- Manual for maintenance
- Spare parts lists

All documents are then checked with regard to completeness, correctness and standards they need to comply with.

It can be seen that in the course of the engineering design process it is essential to communicate to others first ideas, concepts and preliminary design modules as well as the final design. The modern world of engineering is not a one-person show. Instead, it is the result of interdisciplinary teamwork. Therefore, providing a clear picture of such a technical idea and conveying it to various people, inside or outside an organisation, is of high importance and requires an effective and efficient mode of communication.

1.2 COMMUNICATION MODES

Since the beginning of humankind, people have developed several modes of communication. One of the first communication modes was sound. A variety of sounds had different meanings, such as warning others from an existing threat. As a drawback, sounds could be interpreted in different ways, if not commonly agreed on in a group. In addition, communication with the help of sound was limited to basic information.

As civilisation improved, humans used signs for communication. This could have been a fire signal, a pattern of stones, or symbols such as the ancient Sumerian cuneiform and the Egyptian hieroglyphics which were developed in 3500–3000 BC. With the development of alphabets, the written language became a versatile and sophisticated mode of communication.

In the same period, humans started to express themselves with the help of more complex sketches and pictures. Graphics language was born. Over the millennia, graphic representations became a basic and natural form of communication, regardless of the spoken language.

In the engineering field, the effectiveness of graphics language can be easily seen when trying to fully describe an object verbally or in writing. While this might be possible for simple objects, a precise verbal or written description without ambiguity of a more complex object seems to be impossible. As an example, see Figure 1.2 which shows a V8 engine block.

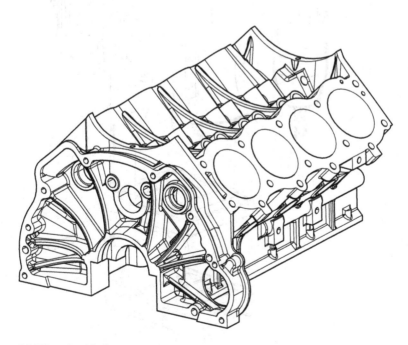

Figure 1.2 V8 engine block.

Graphic representations can be divided into artistic drawings, technical drawings and (technical) illustrations. Drawings from artists are a form of graphic expression of feelings or an idea and have appreciation as their focus. The purpose of an artistic drawing frequently is aesthetic. This kind of drawing is usually subjectively interpreted. In contrast, technical drawings have the purpose of conveying clear information about an object or system, especially how it is constructed or how it functions. As a result, technical drawings are not subject to interpretation but have an intended meaning with no room for misinterpretation.

While drawings and illustrations are both visual representations to convey a particular message by its creator, an illustration is neither purely artistic nor purely technical. Instead, an illustration usually supports the understanding of an accompanying textual content. Within the engineer's world, this kind of graphic representation typically can be found in product documentation such as user manuals, operator manuals or maintenance manuals, for example for machines, automobiles or consumer products. As an example, Figure 1.3 shows a technical illustration of a pillar drill.

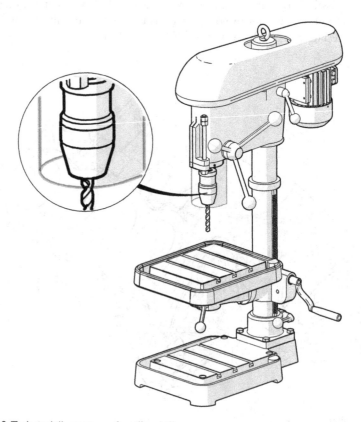

Figure 1.3 Technical illustration of a pillar drill.

Technical illustrations enable us to transfer product-related information or to convey complex matters. When communicating with the general public, this type of graphic representation can enhance the interest and understanding of a non-technical audience. Simplification of a product or assembly usually helps to draw attention to the area, feature or component of interest. For that reason, the illustration should resemble the product in question as much as possible but should also omit details which do not contribute to understanding.

The reason that illustrations are so much better than a mere verbal description is that text always needs to be processed in a sequential manner. Illustrations can be analysed within an instant and need no translation. In the end, the saying 'A picture is worth a thousand words' also applies to technical applications.

While technical illustrations can be the optimum communication mode for certain applications, in some cases a pure graphical representation is not precise enough and needs additional information, especially when it comes to communication on a more scientific level or between experts in the engineering field. So the graphic representation would need to be enriched by specialised technical terminology and symbols. This brings us to technical drawings. In contrast to artistic drawings and technical illustrations, this kind of graphic representation is a detailed and precise document that conveys information about how an object is designed and functions, and how it is supposed to be manufactured and assembled. The drawing is the road map which shows how a product functions, how it had been designed or how it is going to be manufactured. To guarantee that a technical drawing can be understood by any engineer, no matter of which origin or educational background, the elements of the graphic representation follow a set of international standards. By applying internationally developed and approved regulations to standardise the language used, ambiguous interpretation can be avoided. Figure 1.4 depicts a technical drawing of the previous example of a pillar drill. When comparing Figure 1.3 and Figure 1.4, one can realise that comprehension of the overall geometry and identification of individual components is much easier in the technical illustration. The technical drawing provides more information, such as the overall dimensions and the geometric relationships between the components due to the sectional view, which also enables identification of internal parts.

Nowadays, technical drawings are created with the help of computer-aided design (CAD) software packages. In the 1970s and 1980s, the main purpose of using a 2D CAD system was to create a technical drawing. With the introduction of 3D CAD systems in the late 1980s and early 1990s, the desired end product became a virtual three-dimensional model which contains a description of the geometry and serves as a common basis for downstream processes such as simulation, analysis and machining. While manual drafting required a large workspace, specific drafting equipment, and a lot of time to create and edit a drawing, computer-based drafting allows accurate and reproducible engineering drawings in a shorter time, since they can be automatically derived from the driving 3D model. As already mentioned, creating a technical drawing nowadays is not the sole advantage of such CAD systems. Further development

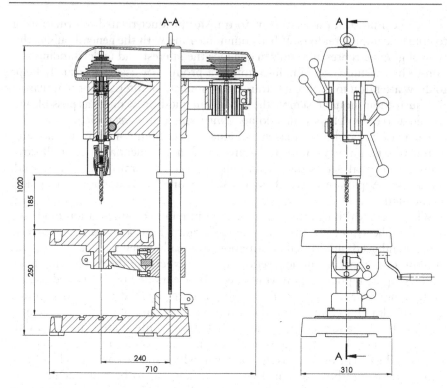

Figure 1.4 Technical drawing of a pillar drill.

of the functionality and increasing computing power led to expert systems for conceptualisation and the design of technical solutions. Regarding communication, 3D CAD systems meanwhile offer various possibilities to represent an object as realistically and as precisely as possible. Figure 1.5 demonstrates how realistic the pillar drill from previous paragraphs can look when a rendered model is created with the help of a modern CAD system.

If CAD systems can provide such impressive photo-realistic images of an object, why do we need (manual) drafting then? It seems that drafting skills are obsolete and that the necessary messages in an engineering design process can be transported via virtual models which show all the characteristics of an object to a very high level of detail.

1.3 IMPORTANCE OF ENGINEERING DRAWING

Engineering drawings still play a major role in the engineering design process since they can be easily understood, once the drafting conventions are known. They can also provide a lot of information regarding a technical solution, beyond the apparently visible. Accompanied by standardised ideograms or annotations, engineering drawings become the ideal mode of communication to

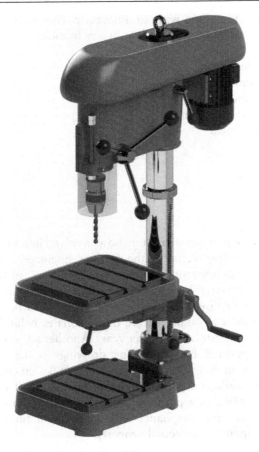

Figure 1.5 Rendered CAD model of a pillar drill.

convey information between the various parties involved in the design process and even the product development process. Owing to their versatile possible use, engineering drawings serve at minimum one of three purposes, frequently all of them at the same time:

- Communication
- Visualisation
- Documentation

As an example, companies use engineering drawings, in either paper or digital form, to convey how to fabricate a part while taking all necessary specifications into account. For that reason, precision needs to be the foremost quality of such a manufacturing drawing. However, communicating a part's manufacturing requirements and specifications is not the sole application. There are numerous interfaces across a company where a design solution needs to be communicated and visualised without ambiguity. This includes the following departments, depending on the specific industry branch:

- Mechanical design
- Electrical design
- Automation and control
- Research and development (R&D)
- Project management
- Purchasing
- Production
- Quality control
- Maintenance
- Service

Usually all these departments are somehow involved in a product's life cycle and therefore need a clear picture of the product. Engineering drawings represent the common communication medium and therefore can be considered to be one of the most important documents in an industrial company.

Once a design is finalised, the design process needs to be diligently documented. One reason is to communicate the rationales behind a design decision, which can be understood even after years. Another advantage of thorough documentation supported by engineering drawings is that future problem statements might be of similar nature so that a new design could be based on existing technical solutions. Apart from the internal usage of such documentation, engineering drawings are also created for legal and archival purposes. They can prove necessary compliance with safety regulations or support the patenting of a company's intellectual property.

Chapter 2

Fundamentals of Engineering Drawing

2.1 INTRODUCTION

As presented in Chapter 1, engineering drawings are never made up of graphics language alone. As a rule, they are a combination of graphics language and written language. Graphics in drawings use lines to represent edges, contours of an object and even surfaces. To distinguish between different types of edges or objects, a variety of different line types and line thicknesses is used. In addition to that, different projection methods help in visualising the object of interest in the best possible way in order to convey the necessary information. The types of projection will be explained in Chapter 4.

Graphics language used in engineering drawings mainly describes a shape and the appearance of an object. On the other hand, written language describes the many details, including the size, location and specification of an object.

From the next section on, an overview is provided of the fundamental elements of graphics and written language, used for any kind of technical drawing.

2.2 DRAFTING EQUIPMENT

2.2.1 Manual Drafting

Basic Tools

Someone who wants to create a technical drawing, such as a draftsperson or engineer, needs some basic tools for drawing. These should include the following:

- A range of pencils
- Rubber
- Ruler
- Set squares
- Protractor
- Compasses
- Dividers
- Drawing board

DOI: 10.1201/9781003001386-2

- T-square
- Clips or tape
- Emery board or fine sandpaper

Pencils: You will need a selection of pencils. They are available with different hardness grades, suitable for different applications. In Europe and most parts of the world, pencil manufacturers use the HB grading system, based on the letters H (hard), B (black = soft) and F (firm). A number indicates the level of hardness. In the USA, the hardness scale is limited to five different grades only, indicated by a number. Figure 2.1 gives an overview of pencil lead hardness scales. For technical drawing purposes, a hard leaded pencil (5H) can be used for light lines, a less hard pencil (2H) for the outlines, and a medium soft pencil (HB) for printing. More than one pencil of each grade will save you from frequent resharpening.

Rubber: Choose a good quality rubber, one that does not smudge.

Ruler: It is advisable to have a transparent ruler. It is also recommended to have a ruler with metric and imperial units (Figure 2.2) since it might happen that you are confronted with different unit systems.

Set squares: You will need at least two set squares: a 60° and a 45° set square (Figure 2.3). It will be also useful to have an adjustable set square, which will enable you to set the angle on the set square to anywhere between 0° and 90°. If you have an adjustable set square, you can manage without the other two.

Protractor: Similar to set squares, the protractor is usually made of transparent acrylic plastic and is helpful for laying out specific angles or measuring an existing angle (Figure 2.4).

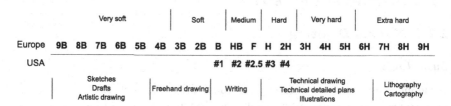

Figure 2.1 Pencil lead grades.

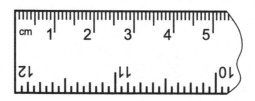

Figure 2.2 Ruler.

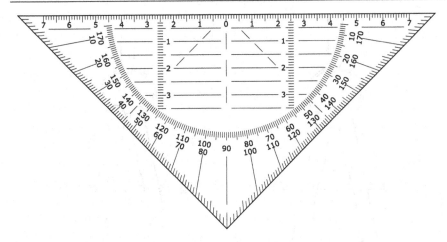

Figure 2.3 Set square (45°).

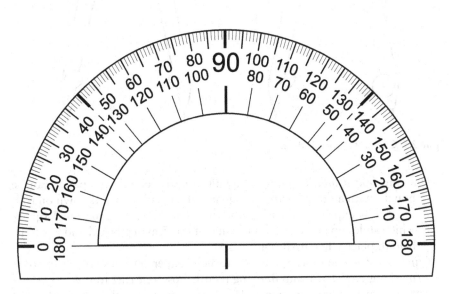

Figure 2.4 Protractor.

Compasses: Circles and arcs in your drawing are created with compasses. You will need at least two compasses: a small spring bow compass (Figure 2.5, left side) for small circles and one for larger circles.

Dividers: Dividers are similar to compasses, except that they are not used for drawing a circle or an arc. Instead, a pair of dividers (Figure 2.5, right side) is very helpful when precisely transferring a distance between views or from one drawing to another. A typical application is dividing a circle or a straight line into equal parts by setting a distance repeatedly. For that reason, both tips of the dividers are pointed.

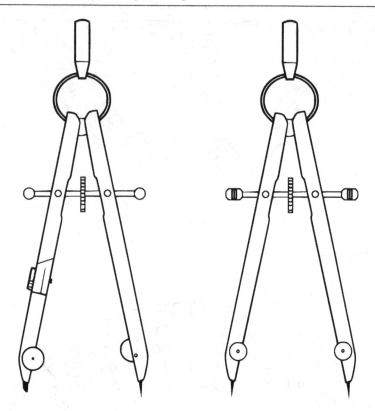

Figure 2.5 Compasses and dividers.

Drawing board and T-square: Drawing boards for size A2 paper can be bought with a fitted horizontal square which slides up and down on rollers. A less expensive board is one that is used with a separate T-square which slides up and down on the side of the drawing board and has to be held in place when used (Figure 2.6).

Clips or tape: The best tape to use to hold paper on the drawing board is masking tape, but metal drawing board clips are easier to use.

Emery board or fine sandpaper: This is used to ensure that the lead in the pencils is kept sharp.

There are other instruments that will help you to draw quickly and accurately. These include the following:

- *French curves* for drawing non-circular and irregular curves accurately (Figure 2.7).
- *Templates* such as circle templates, branch-specific symbol templates and templates with commonly used shapes (Figure 2.8).
- *Scales* (rulers with special markings for drawing items bigger or smaller than they are in real life).

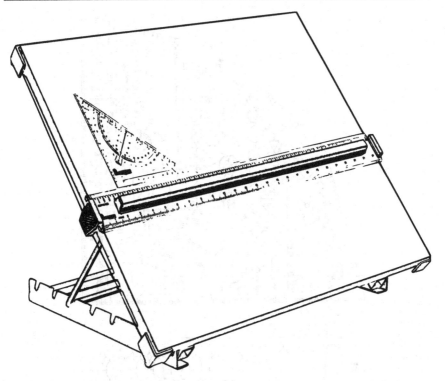

Figure 2.6 Drawing board with T-square and set square.

Figure 2.7 French curves.

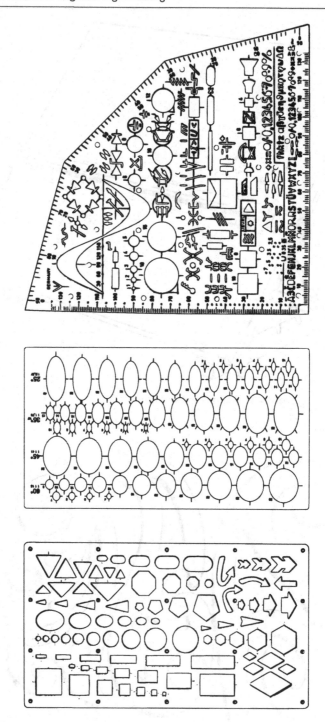

Figure 2.8 Templates.

Using the Equipment

First, fix the paper to your board, using clips or tape. Then sharpen your pencils, either to a point using a pencil sharpener or to a chisel shape using the emery board and use this shape for drawing lines, drawing from the ends of each line to meet in the middle. Use the emery board to sharpen your compass leads to a chisel point too. Finally, draw a frame on your paper if required (see Figure 2.20). Now you are ready to start drawing. Here are some exercises.

First construct an equilateral triangle (Figure 2.9).

Then find the centre of the circumscribing circle.

Draw the circumscribing circle.

Practice shading some parts with the 60° set square.

First draw a circle and step the radius around it six times, starting at the top (Figure 2.10).

Draw the regular hexagon.

Shade the parts of the hexagon with the 60° set square.

2.2.2 Computer-Aided Design (CAD)

Recent decades have been characterised by an accelerating pace in the development of computer technologies. Naturally, this advancement also affected engineering activities, especially the design process and how technical drawings are created. Software solutions for supporting such computer-aided design

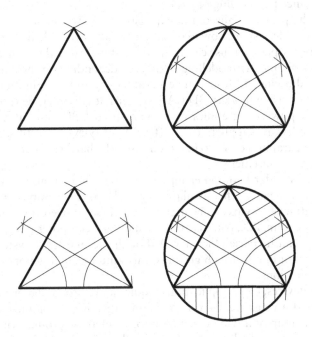

Figure 2.9 Circumscribing circle of an equilateral triangle.

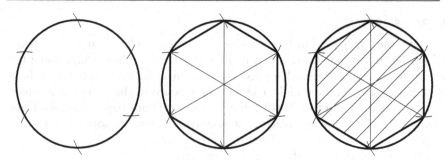

Figure 2.10 Regular hexagon.

process already exist since beginning of the 1960s. While at the beginning these IT solutions were very limited in their functionality, especially in the 1980s and 1990s CAD evolved into a very sophisticated tool to create legible and accurate technical drawings at considerable speed. With comprehensive advent of 3D CAD systems in the 1990s, companies were able to increase their productivity by creating virtual 3D models which can be used for numerous downstream engineering processes such as drafting, simulation and manufacturing. Nowadays, 3D CAD systems are considered to be so-called product lifecycle systems since they take different aspects from all stages of a product's lifecycle into account, i.e. from design through testing, production, commissioning, end use and maintenance to recycling or disposal.

While in the past specialised and expensive CAD workstations were necessary to run a CAD system, nowadays commercially available desktop computers or laptops for personal use can be used for standard CAD applications such as drafting or simple 3D modelling. However, depending on the complexity of the technical drawing, drafting can be a graphically and computationally intensive task. When using modern 3D CAD systems, it may even be recommended to consider a powerful graphics card, high-speed CPU and sufficient RAM. Most CAD software suppliers even offer an overview of certified, i.e. tested and approved, graphics cards to make sure that the hardware complies with the system-specific requirements.

Apart from standard computer input devices such as mouse, keyboard and scanner, there are many other available for CAD. As an example, graphics tablets with an attached stylus can be used to ease cursor movements and the selection of menu commands. Alternatively to standard computer mice, a trackball or 3D mouse can increase efficiency. Although the first CAD systems already support touch and multi-touch gestures, this kind of input is not yet standard for such applications.

As output devices, CAD applications require a monitor and a printer or plotter to put the drawings created on screen on paper. It is recommended to use a dual-monitor setup or a widescreen monitor to increase productivity.

Despite the potential advantage of increased efficiency by using modern CAD systems, manual drafting skills are still important, especially for sketching, quickly communicating and visualising an idea and for understanding the techniques behind computer-aided drafting commands.

2.3 DRAFTING CONVENTIONS

The first step a manufacturer must take when they intend to produce goods is to create a drawing. First, a designer will make a preliminary sketch and then a draftsperson or engineer will make a detailed drawing of the design. Since neither the designer nor the draftsperson will actually manufacture the product, the drawings must be capable of being interpreted by people in the workshops. These workshops may be sited a long way from the drawing office, possibly even overseas, so the drawings produced must be standardised so that anyone familiar with these standards could make the product required, independent of the local spoken language. This is why you will often see symbols used on a drawing instead of words or abbreviations.

National as well as international standards specify the rules for engineering drawing and should be carefully studied by every prospective engineer and draftsperson.

The following chapters explain, within the international framework of ISO 128 (*Technical product documentation and specification*), the language of engineering drawing.

2.3.1 Standardisation

To create clear, coherent and comprehensible drawings, an agreement regarding the commonly used graphics language and written language is required. This is mainly to make sure that a drawing can be understood by everyone without ambiguity. The effectiveness of such drafting language emerged due to the establishment of standards and conventions based on best practices. While the latter is considered to be an instruction set on how to create and read drawings, we use methods of descriptive geometry to implement graphics language. The consistent application of standards in the drafting process guarantees that all manufacturers, or others involved in the product development process, interpret the drawing in the same way. This is a fundamental requirement for interchangeability of parts. Modern industry has been developed on the basis of interchangeable manufacturing and therefore it requires national as well as international standards.

Written or spoken language had been continuously standardised over centuries and is still an ongoing process. The latest status of a standardised language can be found in a dictionary. Likewise, engineering standards are regularly revised and continuously updated. National as well as international organisations make sure that best practices are turned into standards for various industries.

With the transition to more elaborate manufacturing processes during the Industrial Revolution in the 19th century, the need for precise machine tools and interchangeable parts arose. Differences in local standards were making trade very difficult, sometimes even impossible in terms of cost. As a consequence, in 1901 the world's first national standards developing organisation was born, the British *Engineering Standards Committee*. After the First World War, numerous countries followed and set up their own national standard bodies. Finally, in 1947 the *International Organization for Standardization* (ISO) was founded by 25 national standards organisations to ease the exchange of goods and reduce trade barriers. Currently, ISO consists of 165 members. In 1961 the *European Committee for Standardization* (CEN) followed, which brings together the national standardisation bodies of 34 European countries. The idea of EN standards is to successively substitute national standards of the member states for European standards. Table 2.1 gives an overview of some standardisation organisations.

Some of the standards developing organisations are supported by national organisations such as the *Association of German Engineers* (VDI) in Germany and the *American Society of Mechanical Engineers* (ASME) in the USA.

Today, we are facing an almost innumerable variety of national and international standards, which complicates the appropriate selection. If not already restricted by requirements and specifications, ISO standards should take absolute precedence over regional (such as EN) or national standards. Within the European Community, EN standards take higher precedence over national standards of their member countries.

Sometimes we can observe a smooth transition from international to national standards. For instance, an international ISO standard can be transformed into an EN standard and finally into a national standard. The following list exemplifies some of the possible combinations:

- DIN xxxx – German standard with mainly national relevance.
- DIN EN xxxx – European standard which had been transformed into a national standard. Whenever European standards are taken over, they need to be adopted by the members of the European Committee for Standardization without any modifications.

Table 2.1 Examples of national and international standardisation organisations

Region/country	Code	Name of developing organisation
World	ISO	International Organization for Standardization
Europe	EN	European Committee for Standardization
Australia	AS	Standards Australia
Germany	DIN	German Institute for Standardization
Japan	JIS	Japanese Industrial Standards Committee
P.R. China	GB	Standardization Administration of China
United Kingdom	BS	British Standards Institution
USA	ANSI	American National Standards Institute

- DIN EN ISO xxxx – International standard, developed by ISO or CEN and published by both organisations, which is transformed into a German national DIN standard.
- DIN ISO xxxx – Directly from ISO adopted standard.

Within this book, we will mainly address international standards (ISO). All ISO standards addressed in this book can be found in Appendix A.

The reader may refer to their own national standards whenever requirements and specifications do not allow the application of international standards.

2.3.2 Line Conventions

Technical drawings usually consist of a variety of types of lines, each of them with a specific meaning. Part 2 of the international standard ISO 128 establishes the different lines, their designations and examples of application. Most commonly known line types are continuous lines, dashed lines and long-dashed dotted lines, each of them available as a narrow or wide line. In fact there are many more line types available for drafting. To get a clear overview, in ISO 128–2 the lines are classified according to the scheme shown in Figure 2.11.

The most important basic line types are shown in Figure 2.12. Note, this is just an extract out of 15 available basic line types and shows lines commonly used in mechanical engineering practice. Note that the long-dashed double-dotted lines are also referred to as *phantom lines*.

The basic types of lines can also vary with regard to their shape. As an example, instead of a straight continuous line, the basic line type 01 can also be shown as a wavy, spiral, zigzag or freehand continuous line.

XX.Y.Z

Basic line type   Sub line type   Line type application

Figure 2.11 Line type designation according to ISO 128–2.

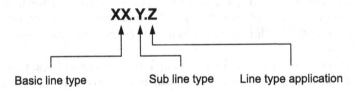

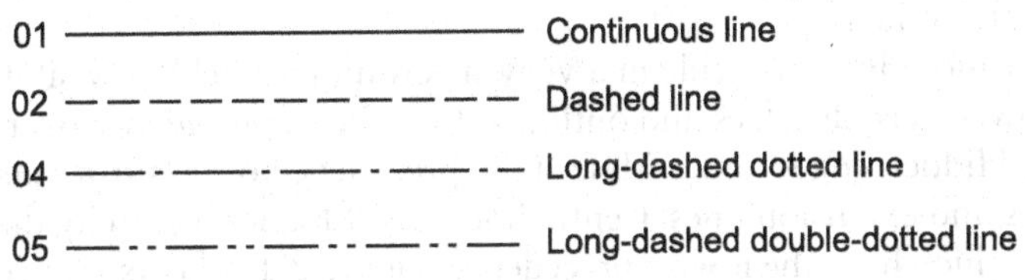

01	——————————	Continuous line
02	— — — — — — — —	Dashed line
04	—— · —— · —— · —— · —	Long-dashed dotted line
05	—— · · —— · · —— · · —— · · ——	Long-dashed double-dotted line

Figure 2.12 Basic line types.

Table 2.2 Line groups and line thickness values

Line group	Wide lines (mm)	Dimensions, callouts, annotations (mm)	Narrow lines (mm)
0.25	0.25	0.18	0.13
0.35	0.35	0.25	0.18
0.5	0.5	0.35	0.25
0.7	0.7	0.5	0.35
1.0	1.0	0.7	0.5
1.4	1.4	1.0	0.7
2.0	2.0	1.4	1.0

As line subtypes, ISO 128–2 distinguishes only between narrow (subtype 1), wide (subtype 2) and extra wide (subtype 3), whereas for most engineering drawings, only two thicknesses are needed, narrow and wide. The final thickness value depends on the selected so-called line group. The width of all types of lines can vary within a predefined series of thickness values, starting from 0.13 mm and ending at 2 mm, with a common ratio between two consecutive values. Table 2.2 gives an overview of the different line groups and the associated line thickness values for different line subtypes.

The line group shall be chosen corresponding to the drawing type, the paper size of the drawing and the applied scale. As a rule, in mechanical engineering applications line group 0.5 is used.

The last part of the line type designation consists of a number indicating the application of a line in technical drawings. As an example, the line type designation 01.1.5 indicates a continuous narrow line, which is used for hatching (application number 5). For details regarding the application number, see ISO 128–2.

In Figure 2.13 and Figure 2.14, different types of lines are exemplarily shown in technical drawing views. For clarity, according to the aforementioned classification, the visible outline is shown as a wide continuous line (basic line type 01, subtype 2).

When it comes to junctions between any non-continuous lines, they should meet at a dash. Otherwise, this could lead to confusion. In case of parallel lines of same line type, the lines should be shown staggered for clarity. However, while in manual drafting you have full control of how to draw any non-continuous line, in CAD systems the exact position of each dash is hard or even impossible to influence.

If two or more lines coincide in a view, a certain hierarchy of visibility needs to be followed. Visible edges and outlines always take precedence over all other line types. Hidden edges and outlines take precedence over cutting plane lines, centre lines and extension lines. Centre lines are of higher hierarchy than extension lines, which have the lowest precedence. Figure 2.15 shows some examples of overlapping lines and their hierarchy of visibility.

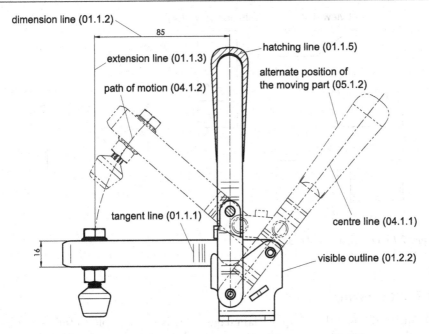

Figure 2.13 Examples of line types in a technical drawing of a toggle press clamp.

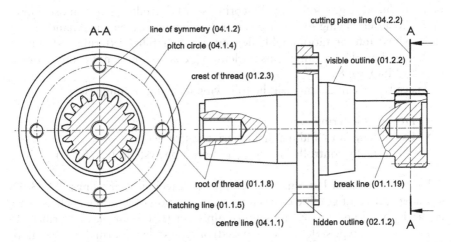

Figure 2.14 Examples of line types in a technical drawing of a tapered shaft.

In case (A) of Figure 2.15, a visible edge coincides with the centre line of the hole. Since the visible edge takes precedence over all other lines, the visible edge is shown and just outside of the outline the centre line is adumbrated. In case (B), a hidden edge and the centre line of the hole fall together. The hidden line is shown since the centre line has a lower precedence. In case (C), a visible edge and a hidden edge coincide.

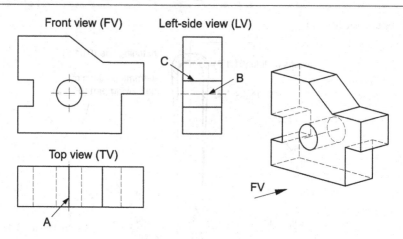

Figure 2.15 Precedence of lines.

2.3.3 Lettering

Owing to the fact that graphics language alone is not suitable to fully describe an object without ambiguity, engineering drawings generally contain some writing in the form of dimensions or notes. Since drawings sometimes need to be reproduced (e.g. photocopy, blueprint, scan), legibility is of utmost importance. To avoid having someone create a drawing using a personal handwriting style, whose interpretation highly depends on readability, lettering is standardised nowadays. The requirements for lettering can be found in the international standard ISO 3098.

Basically, we can distinguish the following lettering techniques:

- Freehand lettering
- Use of templates and manual lettering instruments
- Numerically controlled lettering and drafting systems (e.g. CAD)

Different styles of lettering can be applied in technical drawings. ISO 3098 considers different graphic character sets, inclination of lettering such as vertical and sloped and finally the kind of alphabet (Latin, Greek or Cyrillic). In most drawings, type style A (close-spaced) or type style B (normal) is used. Both styles can be drawn vertical or italic, i.e. slanted to the right by 15°. Vertical lettering style B is recommended by ISO 3098 for standard applications. Figure 2.16 shows a sample of Latin characters with lettering style A.

The height of upper case letters depends on the used line group. Since usually line group 0.5 is used, the line thickness for lettering is 0.35 mm (compare Table 2.2). Character height equals line thickness multiplied by 10 and therefore results in 3.5 mm.

Many draftspersons develop great skills in printing by hand. If you need to print manually, try both standard and italic and develop a style that suits you.

Figure 2.16 Lettering style A (vertical and italic) according ISO 3098.

2.3.4 Drawing Layout

Most paper comes in standard sizes. Metric sheet sizes are internationally standardised in ISO 216. The underlying principle is to have a basic size A0 and to derive the next smaller sheet size by halving (compare Figure 2.17). This leads to a set of geometrically similar sheet sizes. ISO 216 distinguishes a normal A series (regularly derived sizes) and an alternative B series.

Similarly, US customary sheet sizes are standardised by ANSI/ASME Y14.1 and range from ANSI A (smallest size) to ANSI E (largest size). In contrast to ISO 216, the aspect ratio is not constant but alternating.

The largest sheet you are likely to use in technical drawing is A0 and the smallest A4. If your drawing paper has no frame, then draw one. A minimum of 20 mm is used on A0 and A1 from the edge of the paper to the frame line and a minimum of 10 mm on A2, A3 and A4.

In order to space out the views that you will draw on your paper, use the following formulas (A, B and C are the maximum sizes of your views) and the p and q dimensions are the distances between the views.

You do not have to use exact dimensions which might complicate the sums; use sensible approximations for A, B and C (Figure 2.18).

You may well have to add information to your finished drawing, and this should be shown in a block. This information could include a drawing title, the drafter's name, the scale of the drawing, details of approval, the system of projection used, and the date of creation.

In order to ensure compatibility when exchanging drawings, data fields of such a title block are standardised in ISO 7200. It is recommended to have a minimum set of information to facilitate the use and also reuse of the technical drawing.

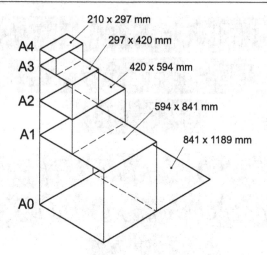

Figure 2.17 Sheet sizes according ISO 216.

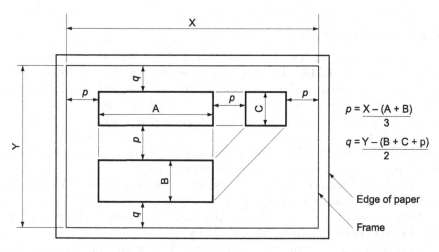

$$p = \frac{X - (A + B)}{3}$$

$$q = \frac{Y - (B + C + p)}{2}$$

Figure 2.18 Positioning of views to be drawn.

Figure 2.19 shows a typical title block which can be found on a technical drawing. Note that the title is given prominence over all the other information. Also note that the system of projection is also given within the title block. This is because assuming the wrong projection method could lead to confusion. Optional data fields are highlighted in Figure 2.19. Although it is not mandatory to provide the data fields shown, it is at least recommended. Finally, it depends on the company size, the company structure and the internal processes.

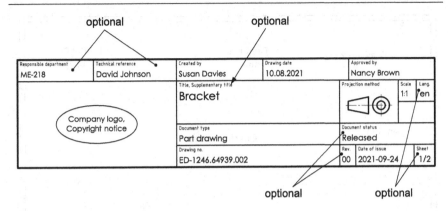

Figure 2.19 Example of a title block.

The same title block is used for all paper sizes. For all landscape oriented paper sizes (usually A0 to A3), the title block is located at the lower right corner of the drawing space. For the size A4 which usually is positioned vertically, the title block is situated in the lower part of the drawing space.

ISO 5457 provides details for drawing sheets such as the available sizes and recommended layouts. The sheet sizes depend on the paper sizes specified in ISO 216, but the available drawing space will be smaller since we have to consider a drawing frame which limits the drawing space and some margins. When a drawing becomes more complex, a grid reference system helps. Locating drawing details such as annotations or revisions will become easier when using a grid of equal zones, overlaying the drawing sheet. ISO 5457 recommends dividing each side of the drawing space into fields of 50 mm length. The fields are vertically referenced with capital letters and horizontally with numbers, located in the grid reference border and originating at the upper left corner (see Figure 2.20).

You might also show an assembly with numerous components on your drawing. For more detailed specification of the individual parts, a parts list can be included in the drawing either in conjunction with the title block (compare Figure 2.20) or placed at any other location. The list of part numbers would be used only if several parts were drawn on the same drawing and would not, therefore, be shown on every drawing.

2.4 SCALES

Before you start any drawing, you first decide how large the drawings have to be. The different views of the object to be drawn must not be bunched together or be too far apart. If you are able to do this and still draw the object in its natural size, then obviously this is best. This is not always possible; the object may be much too large for the paper or much too small to be drawn clearly. In either case it will be necessary to draw the object 'to scale'. The scale must

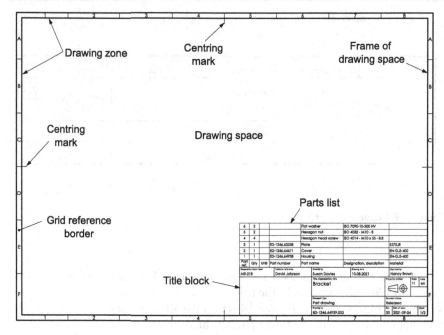

Figure 2.20 Drawing layout with grid reference system.

depend on the size of the object; a miniature electronic component may have to be drawn 100 times larger than it really is, whilst some maps have natural dimensions divided by millions.

There are drawing aids called 'scales' which are designed to help the draftsperson cope with these scaled dimensions. They look like an ordinary ruler, but closer inspection shows that the divisions on these scales are not the usual centimetres or millimetres, but can represent them. These scales are very useful, but there will come a time when you will want to draw to a size that is not on one of these scales. You could work out the scaled size for every dimension on the drawing but this can be a long and tedious business – unless you construct your own scale. This chapter shows you how to construct any scale that you wish.

The Representative Fraction (RF)

The RF shows instantly the ratio of the size of the line on your drawing and the natural size. The ratio of numerator to denominator of the fraction is the ratio of drawn size to natural size. Thus, an RF of $\dfrac{1}{5}$ or 1:5 means that the actual size of the object is five times the size of the drawing of that object.

If a scale is given as 1 mm = 1 m, then the RF is

$$\frac{1\,mm}{1\,m} = \frac{1\,mm}{1000\,mm} = \frac{1}{1000}$$

Plain Scales

There are two types of scale, plain and diagonal. The plain scale is used for simple scales, scales that do not have many subdivisions.

When constructing any scale, the first thing to decide is the length of the scale. The obvious length is a little longer than the longest dimension on the drawing. Figure 2.21 shows a very simple scale of 20 mm = 100 mm. The largest natural dimension is 500 mm, so the total length of the scale is $\frac{500}{5}$ mm or 100 mm. This 100 mm is divided into five equal portions, each portion representing 100 mm. The first 100 mm is then divided into 10 equal portions, each portion representing 10 mm. These divisions are then clearly marked to show what each portion represents.

Diagonal Scales

There is a limit to the number of divisions that can be constructed on a plain scale. Try to divide 10 mm into 50 parts; you will find that it is almost impossible. The architect, cartographer and surveyor all have the problem of having to subdivide into smaller units than a plain scale allows. A diagonal scale allows you to divide into smaller units.

Before looking at any particular diagonal scale, let us first look at the underlying principle.

Figure 2.22 shows a triangle ABC. Suppose that AB is 10 mm long and BC is divided into 10 equal parts. Lines from these equal parts have been drawn parallel to AB and numbered from 1 to 10.

It should be obvious that the line 5–5 is half the length of AB. Similarly, the line 1–1 is $\frac{1}{10}$ the length of AB and line 7–7 is $\frac{7}{10}$ the length of AB. (If you wish to prove this mathematically use similar triangles.)

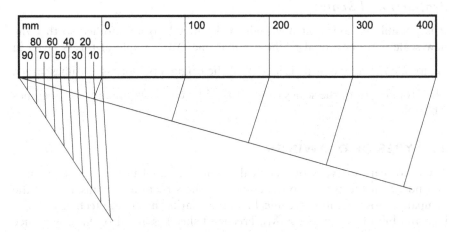

Figure 2.21 Plain scale 20 mm = 100 mm or 1 mm = 5 mm.

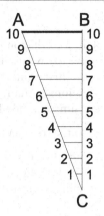

Figure 2.22 Showing how to divide a line AB into 10 equal parts.

You can see that the lengths of the lines 1–1 to 10–10 increase by 1 mm each time you go up a line. If the length of AB had been 1 mm to begin with, the increases would have been $\frac{1}{10}$ mm each time. In this way, small lengths can be divided into very much smaller lengths and can be easily picked out.

An example of diagonal scales follows.

This scale would be used where the drawing is twice the size of the natural object and the draftsperson has to be able to measure on a scale accurate to 0.1 mm.

The longest natural dimension is 60 mm. This length is first divided into six 10 mm intervals. The first 10 mm is then divided into 10 parts, each 1 mm wide (scaled). Each of these 1 mm intervals is divided with a diagonal into 10 more equal parts (Figure 2.23).

Proportional Scales

It is possible to construct one plain scale directly from another, so that the new scale is proportional to the original one. An example of this is given in Figure 2.24. The new scale is a copy of the original one but is $\frac{7}{4}$ times larger.

The proportions of the scales can be varied by changing the ratios of the lines AB to BC.

2.5 TYPES OF DRAWINGS

There are various types of technical documents used in a company, mainly dependent on the type of production and the structural organisation of the company. Most common technical communication media are technical drawings and bill of materials (BOM). Freehand sketches are also important since they can contribute to solution finding, serve as a basis for a discussion or are necessary for measurements on site.

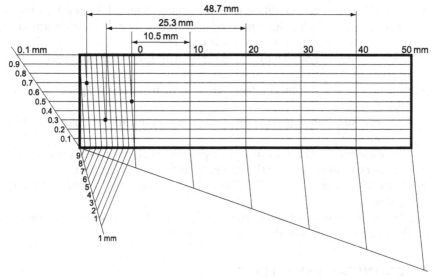

Figure 2.23 Diagonal scale 20 mm = 10 mm to read to 0.1 mm (RF = $\frac{2}{1}$).

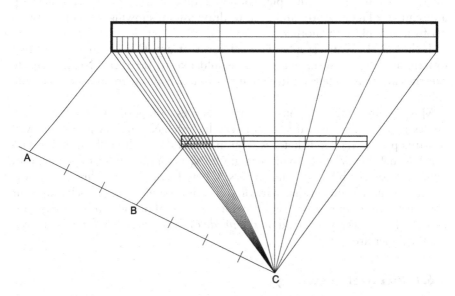

Figure 2.24 To enlarge a scale by a proportion of 7:4.

According to ISO 10209, the following types of technical drawings exist, listed in alphabetical order:

- *Arrangement drawing*: Shows parts and their position relative to each other. This drawing type is used for larger machines and installations.

- *Assembly drawing*: Identifies all components from a high-level group of assembled parts. This drawing type also shows the relative position and/ or shape of the parts as well as their quantity.
- *CAD drawing*: Created with a CAD system and shows the full CAD model or parts of it as presented on screen or printed on paper.
- *Chart*: Provides graphical information in the form of a graph, diagram or table.
- *Dimensional drawing*: Specifies dimensions necessary for manufacturing or mounting.
- *Part drawing* or *detail drawing*: Depicts a single part and provides the necessary information for the complete definition of the part.
- *Production drawing*: Gives all information about a part required for its production.
- *Sketch*: Prepared freehand or with the help of a CAD system, not necessarily to scale or showing the right proportions.

2.6 TECHNICAL SKETCHING

The ability to sketch neatly and accurately is one of the most useful attributes that an engineer or a draftsperson can have. Freehand drawing is done on many occasions: to explain a piece of design quickly to a colleague; to develop a design (see Figure 2.30); and even to draw a map showing someone how to get from one place to another.

Technical sketching is a disciplined form of art. Objects must be drawn exactly as they are seen, not as one would like to see them. Neat, accurate sketches are only achieved after plenty of practice, but there are some guiding rules.

Most engineering components have outlines composed of straight lines, circles and circular arcs: if you can sketch these accurately, you are halfway towards producing good sketches. You may find the method illustrated in Figure 2.25 helpful. When drawing straight lines, as on the left, rest the weight of your hand on the backs of your fingers. When drawing curved lines, as on the right, rest the weight on that part of your hand between the knuckle of your little finger and your wrist. This provides a pivot about which to swing your pencil. Always keep your hand on the *inside* of the curve, even if it means moving the paper around.

2.6.1 Pictorial Sketching

Freehand pictorial sketching looks very much like isometric drawing. Circles appear as ellipses and lines are drawn at approximately 30°. Circles have been sketched onto an isometric cube in Figure 2.26. You can see how these same ellipses appear on sketches of a round bar material.

Isometric drawing will be explained in detail in Chapter 4. For the moment, it is sufficient to know that an isometric view is one approach to represent a three-dimensional object in two dimensions, i.e. on a two-dimensional plane.

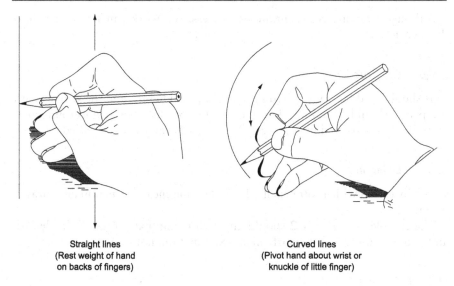

Figure 2.25 Drawing straight and curved lines.

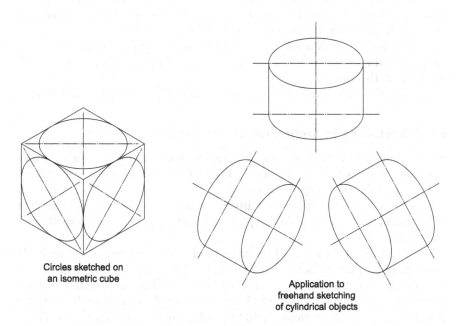

Figure 2.26 Freehand pictorial sketching of circles.

When sketching, you may find it an advantage to draw a faint 'box' first and draw in the ellipses afterwards. With practice you should find that you can draw quite a good ellipse if you mark out its centre lines and the major and minor axes.

Although drawing is a continuous process, the work can be divided into three basic stages.

Stage 1: Construction

This should be done with a hard pencil (6H), used lightly, and the strokes with the pencil should be rapid. Slow movements produce wavy, uncertain lines. Since these constructed lines are very faint, errors can easily be erased.

Stage 2: Lining in

Carefully line in with a soft pencil (HB), following the construction lines drawn in stage 1.

The completion of stage 2 should give a drawing that shows all the details, and you may decide, particularly in an examination, not to proceed to stage 3.

Stage 3: Shading

Shading brings a drawing to life. It is not necessary on most sketches, and in some cases it may tend to hide details that need to be seen. If the drawings are to be displayed, however, some shading should certainly be done.

Shading is done with a soft pencil (HB). It is very easy to overshade, so be careful. For the smooth merging of shading, the dry tip of a finger can be gently rubbed over the area.

Figure 2.27 and Figure 2.28 are examples of freehand pictorial sketching.

2.6.2 Sketching in Orthographic Projection

More detail can be seen on an orthographic drawing than on an isometric, mainly because more than one view is drawn. For this reason it is often advantageous to make an orthographic sketch.

Note, if you are not familiar with the different projection types such as orthographic projection, go to Chapter 4 first which provides a detailed introduction into this topic.

For sketching in orthographic projection, the views should be drawn in the conventional orthographic positions, i.e. in third angle projection, a front elevation (FE), a plan above the FE, and an end elevation (EE) to the left or right of the FE; in first angle projection, an FE, a plan below the FE and an EE to the left or right of the FE. These views should be linked together with projection lines.

Figure 2.29 shows a pictorial and an orthographic sketch of an engineering component.

Circles are difficult to draw freehand but you can use your hand as a compass. Hold your pencil upright and, using your little finger as a compass 'point', rotate the paper keeping your hand quite still.

Figure 2.30 shows how a draftsperson or engineer might use sketching to aid a piece of design. They wish to design a small hand vice.

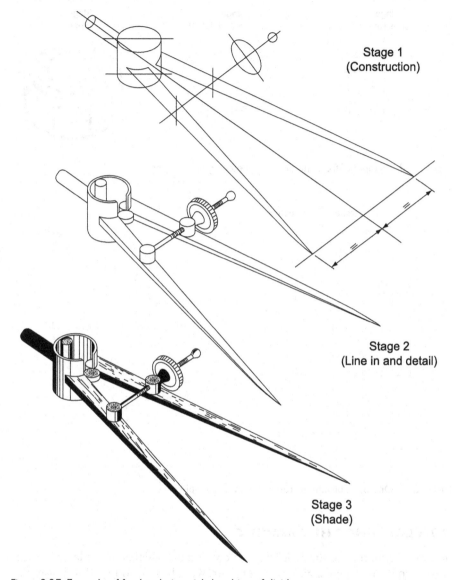

Stage 1
(Construction)

Stage 2
(Line in and detail)

Stage 3
(Shade)

Figure 2.27 Example of freehand pictorial sketching of dividers.

First, they make a freehand pictorial sketch so that they can see what the vice will look like. They also make a few notes about some details of the vice.

They then make an orthographic sketch. This shows much more detail and they make some more notes.

This is quite a simple vice and they may now feel that they are ready to make the detailed drawings. If it was more complicated they might make a few more sketches showing even more detail. Details of one of the legs of the vice are shown.

Stage 1
(Construction)

Stage 2
(Line in and detail)

Stage 3
(Shade)

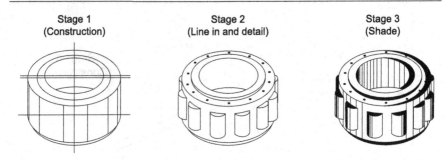

Figure 2.28 Example of freehand pictorial sketching of a bearing.

First angle projection

Section X–X

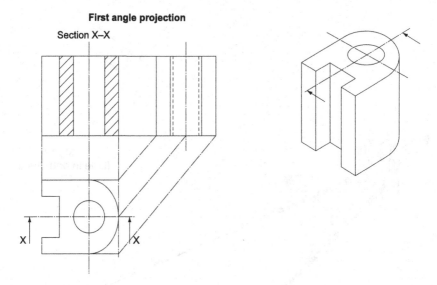

Figure 2.29 Orthographic and pictorial sketch of a simple part.

2.7 PLANNING THE DRAWING

Before beginning the drawings for a new design, whether you plan to use a CAD program or pencil and paper on a drawing board, you need to know the answers to key questions. So ask yourself the following:

(i) Who is the finished article for?
(ii) What is it for?
(iii) What do you want the end result to look like?

Now you can begin to ask yourself some more questions about details:

(a) What approximate size will it be?
(b) What will it be made of?

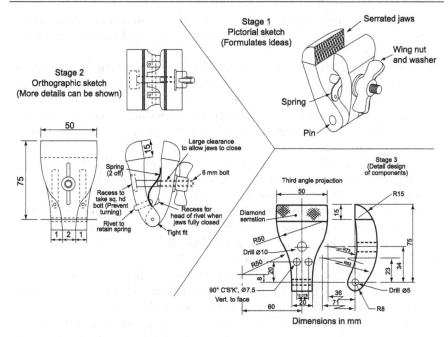

Figure 2.30 Sketching and design.

(c) How many parts will there be and how will you connect them together?

(d) What materials will be required?

(e) What finish will it have?

(f) Will it be packaged, and how much protection is needed with the packaging, i.e. how strong should it be?

Each of these questions raises a new set of questions, and this book should help you to answer some of them. For instance:

- If you are designing a very large or very small object then section 2.4, on scales, might help you with your plan.
- If some of the shapes in your design are likely to contain geometric profiles, then look at section 3.1.
- If the people who you are designing for would appreciate a pictorial view or views of your ideas, then look at Chapter 4 for ways of achieving this.
- If any part or parts of your design involve a whole or part of a circle or circles (wheels, gears, etc.), then check section 3.1.
- If any part or parts of your design use belts, chains or straps, then see if anything in section 3.2 can help you.
- Making figures of all shapes bigger or smaller can be achieved with the techniques shown in section 5.2.

- The outlines of designed objects are rarely composed of just straight lines or whole circles, and ways that these can be blended together are shown in section 3.2.
- The tracks taken by moving parts often need to be checked to ensure that parts do not jam against each other. Techniques to plot these tracks, known as loci, are shown in section 3.3.
- It is often important to show orthographic views other than the two main elevations and the plan. Auxiliary elevations and the way to draw them is shown in section 6.2.
- Shapes like parabolas (seen on suspension bridges and light reflectors), ellipses and hyperbolas are all found by taking sections through a cone. How you do this is shown in section 4.7.
- When two or more geometric shapes run into each other (like a cylindrical pipe running into a cone), it can be useful to know what the line of intersection looks like. A CAD program will do this for you, but if you want to work it out for yourself look at section 5.3.
- Orthographic drawings of three-dimensional figures have to be done on two-dimensional flat surfaces. If an item being drawn involves shapes that are not parallel to any of the three basic drawing planes, then they are distorted. A CAD program might illustrate the true length of a line or the shape of a surface that is not horizontal or vertical. However, if you need to work this out for yourself, then section 5.1 will show you how to do it.
- If the object you design has to be packaged, then you might have to design the packaging – section 5.4 shows you how to find the flat (or developed) shape.
- More examples of loci, particularly the path of parts of rotating wheels and a special path traced out by spirals, are shown in section 3.3.
- Some basic sketching ideas are shown in section 2.6.
- A method to calculate the areas of irregular shapes and a method to resolve some problems about simple forces of the kind met in structural designs and cams design (used inside many engines) are shown in Chapter 12.

2.8 PROBLEMS

(All questions originally set in imperial units.)

Scales

1

(a) Draw the simple key shown in Figure 2.31 full size.

(b) Construct a plain scale with an RF of $\frac{5}{4}$ suitable for use in the making of an enlarged drawing of this key. Do *not* draw this key again.

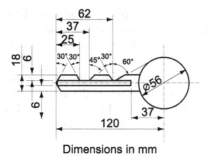

Figure 2.31 Drawing of a key.

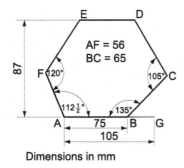

Dimensions in mm

Figure 2.32 Drawing of a polygon.

Southern Regional Examinations Board

2 Construct a plain scale of 50 mm = 300 mm to read to 10 mm up to 1200 mm. Using this scale, draw to scale a triangle having a perimeter of 1200 mm and having sides in the ratio 3:4:6. Print neatly along each side the length to the nearest 10 mm. *Oxford Local Examinations* (See section 3.1 for information not in Chapter 2.)

3 Construct the plain figure shown in Figure 2.32 and then, by means of a proportional scale, draw a similar figure standing on the base AG. All angles must be constructed geometrically in the first figure. Measure and state the length of the side corresponding with CD. *Oxford Local Examinations* (See section 3.1 for information not in Chapter 2.)

4 Construct a diagonal scale in which 40 mm represents 1 m. The scale is to read down to 10 mm and is to cover a range of 5 m. Mark off a distance of 4 m 780 mm.

5 Construct a diagonal scale of 25 mm to represent 1 m which can be used to measure m and 10 mm up to 8 m. Using this scale construct a quadrilateral ABCD which stands on a base AB of length 4 m 720 mm and having BC = 3 m 530 mm, AD = 4 m 170 mm, ∠ABC = 120° and

∠ADC = 90°. Measure and state the lengths of the two diagonals and the perpendicular height, all correct to the nearest 10 mm. Angles must be constructed geometrically. *Oxford Local Examinations* (See section 3.1 for information not in Chapter 2.)

6   Construct a diagonal scale, 10 times full size, to show mm and tenths of a mm and to read to a maximum of 20 mm. Using the scale, construct a triangle ABC with AB 17.4 mm, BC 13.8 mm and AC 11 mm.
Oxford and Cambridge Schools Examinations Board

Chapter 3

Geometric Constructions

3.1 THE CONSTRUCTION OF GEOMETRIC FIGURES FROM GIVEN DATA

This section is concerned with the construction of plane geometric figures. Plane geometry is the geometry of figures that are two-dimensional, i.e. figures that have only length and breadth. Solid geometry is the geometry of three-dimensional figures.

There are an endless number of plane figures, but we will concern ourselves only with the more common ones – the triangle, the quadrilateral and the better-known regular polygons.

Before we look at any particular figure, there are a few constructions that must be revised (Figure 3.1–Figure 3.12).

3.1.1 The Triangle

Definitions

The triangle is a plane figure bounded by three straight sides.

A *scalene* triangle is a triangle with three unequal sides and three unequal angles.

An *isosceles* triangle is a triangle with two sides, and hence two angles, equal.

An *equilateral* triangle is a triangle with all the sides, and hence all the angles, equal.

A *right-angled* triangle is a triangle containing one right angle. The side opposite the right angle is called the 'hypotenuse'.

Constructions

To construct an equilateral triangle, given one of the sides (Figure 3.13).

1 Draw a line AB, equal to the length of the side.
2 With compass point on A and radius AB, draw an arc as shown.
3 With compass point on B, and with the same radius, draw another arc to cut the first arc at C.

DOI: 10.1201/9781003001386-3

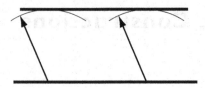

Figure 3.1 To construct a parallel line.

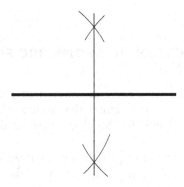

Figure 3.2 To bisect a line.

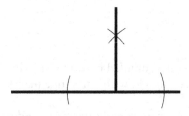

Figure 3.3 To erect a perpendicular from a point on a line.

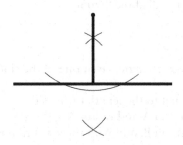

Figure 3.4 To erect a perpendicular from a point to a line.

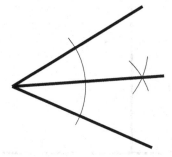

Figure 3.5 To bisect an angle.

Figure 3.6 To bisect the angle formed by two converging lines.

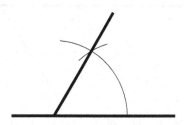

Figure 3.7 To construct 60°.

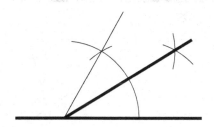

Figure 3.8 To construct 30°.

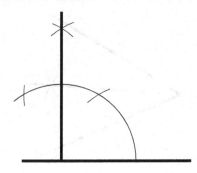

Figure 3.9 To construct 90°.

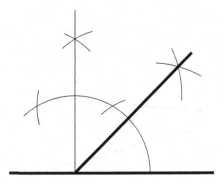

Figure 3.10 To construct 45°.

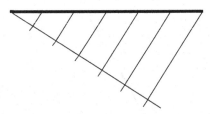

Figure 3.11 To divide a line into a number of equal parts (e.g. six).

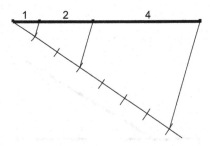

Figure 3.12 To divide a line proportionally (e.g. 1:2:4).

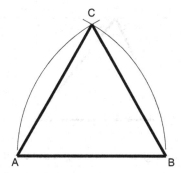

Figure 3.13 To construct an equilateral triangle.

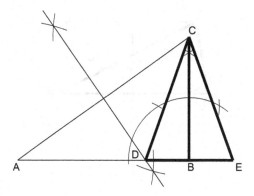

Figure 3.14 To construct an isosceles triangle.

Triangle ABC is equilateral.

To construct an isosceles triangle given the perimeter and the altitude (Figure 3.14).

1 Draw line AB equal to half the perimeter.
2 From B erect a perpendicular and make BC equal to the altitude.
3 Join AC and bisect it to cut AB in D.
4 Produce DB so that BE = BD. CDE is the required triangle.

To construct a triangle, given the base angles and the altitude (Figure 3.15).

1 Draw a line AB.
2 Construct CD parallel to AB so that the distance between them is equal to the altitude.
3 From any point E, on CD, draw CÊF and DÊG so that they cut AB in F and G, respectively.

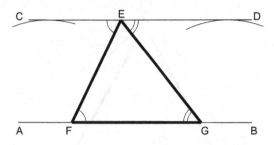

Figure 3.15 To construct a triangle, given the base angles and the altitude.

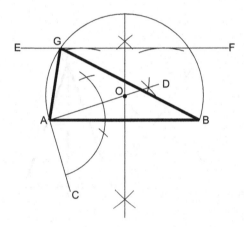

Figure 3.16 To construct a triangle given the base, the altitude and the vertical angle.

Since CÊF = EF̂G and DÊG = EĜF (alternate angles), then EFG is the required triangle.

To construct a triangle given the base, the altitude and the vertical angle (Figure 3.16).

1   Draw the base AB.
2   Construct BÂC equal to the vertical angle.
3   Erect AD perpendicular to AC.
4   Bisect AB to meet AD in O.
5   With centre O and radius OA (= OB), draw a circle.
6   Construct EF parallel to AB so that the distance between them is equal to the altitude.

Let EF intersect the circle in G.
ABG is the required triangle.

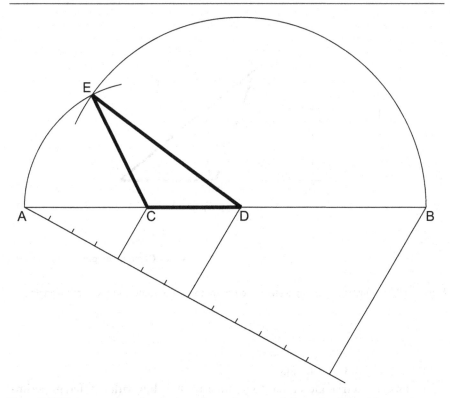

Figure 3.17 To construct a triangle given the perimeter and the ratio of the sides.

To construct a triangle given the perimeter and the ratio of the sides (Figure 3.17).

1 Draw the line AB equal in length to the perimeter.
2 Divide AB into the required ratio (e.g. 4:3:6).
3 With centre C and radius CA, draw an arc.
4 With centre D and radius DB, draw an arc to intersect the first arc in E.

ECD is the required triangle.

To construct a triangle given the perimeter, the altitude and the vertical angle (Figure 3.18).

1 Draw AB and AC each equal to half the perimeter, so that CAB is the vertical angle.
2 From B and C erect perpendiculars to meet in D.
3 With centre D, draw a circle, radius DB (= DC).
4 With centre A and radius equal to the altitude, draw an arc.
5 Construct the common tangent between the circle and the arc. Let this tangent intersect AC in F and AB in E.

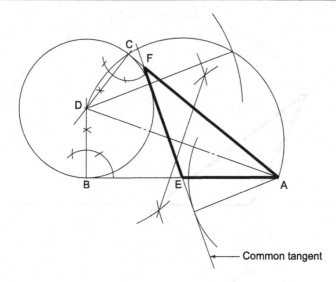

Common tangent

Figure 3.18 To construct a triangle given the perimeter, the altitude and the vertical angle.

FEA is the required triangle.

To construct a triangle similar to another triangle but with a different perimeter (Figure 3.19).

1   Draw the given triangle ABC.
2   Produce BC in both directions.
3   With compass point on B and radius BA, draw an arc to cut CB produced in F.
4   With compass point on C and radius CA, draw an arc to cut BC produced in E.
5   Draw a line FG equal in length to the required perimeter.
6   Join EG and draw CJ and BH parallel to it.
7   With centre H and radius HF, draw an arc.
8   With centre J and radius JG, draw another arc to intersect the first arc in K.

HKJ is the required triangle.

3.1.2 The Quadrilateral

Definitions

The quadrilateral is a plane figure bounded by four straight sides.

A *square* is a quadrilateral with all four sides of equal length and one of its angles (and hence the other three) a right angle.

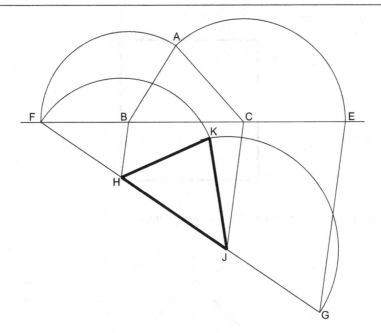

Figure 3.19 To construct a triangle similar to another triangle but with a different perimeter.

A *rectangle* is a quadrilateral with its opposite sides of equal length and one of its angles (and hence the other three) a right angle.

A *parallelogram* is a quadrilateral with opposite sides equal and therefore parallel.

A *rhombus* is a quadrilateral with all four sides equal.

A *trapezium* is a quadrilateral with one pair of opposite sides parallel.

A *trapezoid* is a quadrilateral with all four sides and angles unequal.

Constructions

To construct a square given the length of the side (Figure 3.20).

1 Draw the side AB.
2 From B erect a perpendicular.
3 Mark off the length of side BC.
4 With centres A and C draw arcs, radius equal to the length of the side of the square, to intersect at D.

ABCD is the required square.

To construct a square given the diagonal (Figure 3.21).

1 Draw the diagonal AC.
2 Bisect AC.

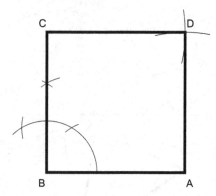

Figure 3.20 To construct a square given the length of the side.

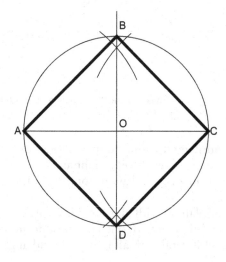

Figure 3.21 To construct a square given the diagonal.

3   With centre O and radius OA (= OC), draw a circle to cut the bisecting line in B and D.

ABCD is the required square.

To construct a rectangle given the length of the diagonal and one of the sides (Figure 3.22).

1   Draw the diagonal BD.
2   Bisect BD.
3   With centre O and radius OB (= OD), draw a circle.
4   With centre B and radius equal to the length of the known side, draw an arc to cut the circle in C.

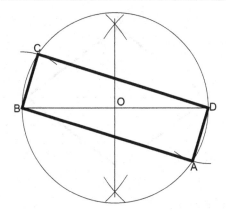

Figure 3.22 To construct a rectangle given the length of the diagonal and one of the sides.

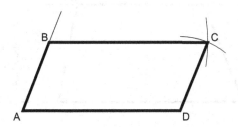

Figure 3.23 To construct a parallelogram given two sides and an angle.

5   Repeat step 4 with centre D to cut at A.

ABCD is the required rectangle.
To construct a parallelogram given two sides and an angle (Figure 3.23).

1   Draw AD equal to the length of one of the sides.
2   From A construct the known angle.
3   Mark off AB equal in length to the other known side.
4   With compass point at B, draw an arc equal in radius to AD.
5   With compass point at D, draw an arc equal in radius to AB.

ABCD is the required parallelogram.
To construct a rhombus given the diagonal and the length of the sides (Figure 3.24).

1   Draw the diagonal AC.
2   From A and C draw intersecting arcs, equal in length to the sides, to meet at B and D.

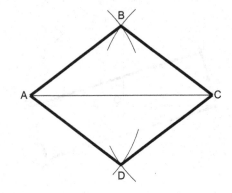

Figure 3.24 To construct a rhombus.

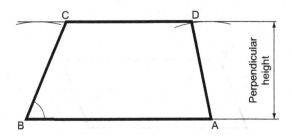

Figure 3.25 To construct a trapezium.

ABCD is the required rhombus.

To construct a trapezium given the lengths of the parallel sides, the perpendicular distance between them and one angle (Figure 3.25).

1 Draw one of the parallels AB.
2 Construct the parallel line.
3 Construct the known angle from B to intersect the parallel line in C.
4 Mark off the known length CD.

ABCD is the required trapezium.

3.1.3 Polygons

Definitions

A polygon is a plane figure bounded by minimum three straight sides. Polygons that are frequently referred to have particular names. Some of these are listed as follows.

A *trigon* (triangle) is a plane figure bounded by three sides.

A *tetragon* (quadrilateral) is a plane figure bounded by four sides.

A *pentagon* is a plane figure bounded by five sides.

A *hexagon* is a plane figure bounded by six sides.

A *heptagon* is a plane figure bounded by seven sides.

An *octagon* is a plane figure bounded by eight sides.

A *nonagon* is a plane figure bounded by nine sides.

A *decagon* is a plane figure bounded by ten sides.

A regular polygon is one that has all its sides equal and therefore all its exterior angles equal and all its interior angles equal.

It is possible to construct a circle within a regular polygon so that all the sides of the polygon are tangential to that circle. The diameter of that circle is called the 'diameter of the polygon'. If the polygon has an even number of sides, the diameter is the distance between two diametrically opposed faces. This dimension is often called the 'across-flats' dimension.

The diagonal of a polygon is the distance from one corner to the corner farthest away from it. If the polygon has an even number of sides, then this distance is the dimension between two diametrically opposed corners.

Constructions

To construct a regular hexagon given the length of the sides (Figure 3.26).

1 Draw a circle, radius equal to the length of the side.
2 From any point on the circumference, step the radius around the circle six times. If your construction is accurate, you will finish at exactly the same place that you started.
3 Connect the six points to form a regular hexagon.

To construct a regular hexagon given the diameter (Figure 3.27).

This construction, using compasses and straight edge only, is quite feasible but is relatively unimportant. What is important is to recognise that a hexagon

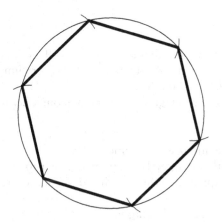

Figure 3.26 To construct a regular hexagon given the length of the sides.

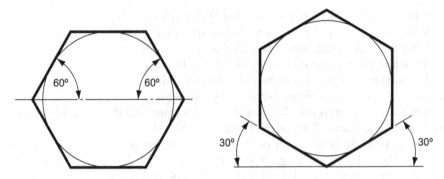

Figure 3.27 To construct a regular hexagon given the diameter.

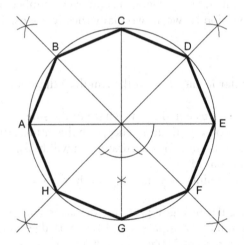

Figure 3.28 To construct a regular octagon given the diagonal.

can be constructed, given the diameter or across-flats dimension, by drawing tangents to the circle with a 60° set square. This is very important when drawing hexagonal-headed nuts and bolts.

To construct a regular octagon given the diagonal, i.e. within a given circle (Figure 3.28).

1   Draw the circle and insert a diameter AE.
2   Construct another diagonal CG, perpendicular to the first diagonal.
3   Bisect the four quadrants thus produced to cut the circle in B, D, F and H.

ABCDEFGH is the required octagon.

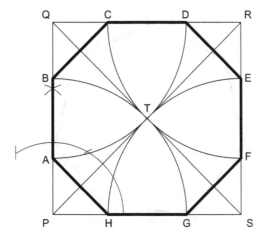

Figure 3.29 To construct a regular octagon given the diameter.

To construct a regular octagon given the diameter, i.e. within a given square (Figure 3.29).

1   Construct a square PQRS, length of side equal to the diameter.
2   Draw the diagonals SQ and PR to intersect in T.
3   With centres P, Q, R and S draw four arcs, radius PT (= QT = RT = ST) to cut the square in A, B, C, D, E, F, G and H.

ABCDEFGH is the required octagon.

To construct any given polygon, given the length of a side.
There are three fairly simple ways of constructing a regular polygon. Two methods require a simple calculation and the third requires very careful construction if it is to be exact. All three methods are shown. The construction works for any polygon, and a heptagon (seven sides) has been chosen to illustrate them.
Method 1 (Figure 3.30).

1   Draw a line AB equal in length to one of the sides and produce AB to P.
2   Calculate the exterior angle of the polygon by dividing 360° by the number of sides. In this case the exterior angle is 360°/7 = 51.4°.
3   Draw the exterior angle PBC so that BC = AB.
4   Bisect AB and BC to intersect in O.
5   Draw a circle, centre O and radius OA (= OB = OC).
6   Step off the sides of the figure from C to D, D to E, etc.

ABCDEFG is the required heptagon.

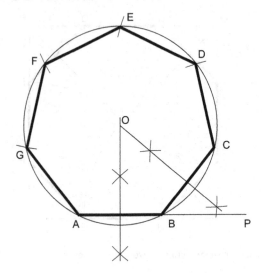

Figure 3.30 To construct a heptagon.

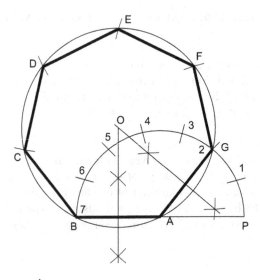

Figure 3.31 To construct a heptagon.

Method 2 (Figure 3.31).

1   Draw a line AB equal in length to one of the sides.
2   From A erect a semi-circle, radius AB to meet BA produced in P.
3   Divide the semi-circle into the same number of equal parts as the pro-
    posed polygon has sides. This may be done by trial and error or by calcu-
    lation (180°/7 = 25.7° for each arc).

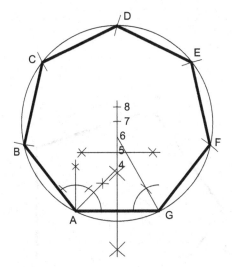

Figure 3.32 Alternative method to construct a heptagon.

4 Draw a line from A to point 2 (for ALL polygons). This forms a second side to the polygon.
5 Bisect AB and A₂ to intersect in O.
6 With centre O draw a circle, radius OB (= OA = O_2).
7 Step off the sides of the figure from B to C, C to D, etc.

ABCDEFG is the required heptagon.
Method 3 (Figure 3.32).

1 Draw a line GA equal in length to one of the sides.
2 Bisect GA.
3 From A construct an angle of 45° to intersect the bisector at point 4.
4 From G construct an angle of 60° to intersect the bisector at point 6.
5 Bisect between points 4 and 6 to give point 5.

Point 4 is the centre of a circle containing a square. Point 5 is the centre of a circle containing a pentagon. Point 6 is the centre of a circle containing a hexagon. By marking off points at similar distances, the centres of circles containing any regular polygon can be obtained.

6 Mark off point 7 so that 6 to 7 = 5 to 6 (= 4 to 5).
7 With centre at point 7 draw a circle, radius 7 to A (= 7 to G).
8 Step off the sides of the figure from A to B, B to C, etc.

ABCDEFG is the required heptagon.

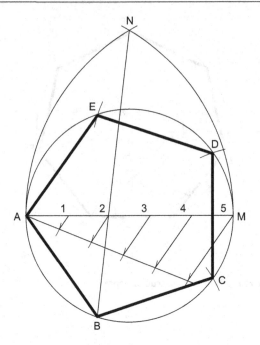

Figure 3.33 To construct a regular polygon given a diagonal.

To construct a regular polygon given a diagonal, i.e. within a given circle (Figure 3.33).

1   Draw the given circle and insert a diameter AM.
2   Divide the diameter into the same number of divisions as the polygon has sides.
3   With centre M draw an arc, radius MA. With centre A draw another arc of the same radius to intersect the first arc in N.
4   Draw N$_2$ and produce to intersect the circle in B (for any polygon).
5   AB is the first side of the polygon. Step out the other sides BC, CD, etc.

ABCDE is the required polygon.
To construct a regular polygon given a diameter (Figure 3.34).

1   Draw a line MN.
2   From some point A on the line, draw a semi-circle of any convenient radius.
3   Divide the semi-circle into the same number of equal sectors as the polygon has sides (in this case nine, i.e. 20° intervals).
4   From A draw radial lines through points 1 to 8.
5   If the polygon has an even number of sides, there is only one diameter passing through A. In this case, bisect the known diameter to give centre

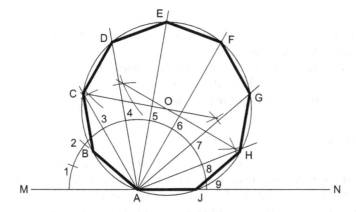

Figure 3.34 To construct a regular polygon given a diameter.

O. If, as in this case, there are two diameters passing through A (there can never be more than two), then bisect both diameters to intersect in O.

6   With centre O and radius OA, draw a circle to intersect the radial lines in C, D, E, F, G and H.

7   From A mark off AB and AJ equal to CD, DE, etc.

ABCDEFGHJ is the required polygon.

The prior constructions shown are by no means all the constructions that you may be required to do, but they are representative of the type that you may meet.

If your geometry needs a little extra practice, it is well worthwhile proving these constructions by Euclidean proofs. A knowledge of some geometric theorems is needed when answering many of the questions shown later, and proving the previous constructions will make sure that you are familiar with them.

3.1.4 Circles

About 6000 years ago, an unknown Mesopotamian made one of the greatest inventions of all time: the wheel. This was the most important practical application ever made of a shape that fascinated early mathematicians. The shape is, of course, the circle. After the wheel had been invented, the Mesopotamians found many more applications for the circle than just for transport. The potter's wheel was developed and vessels were made much more accurately and quickly. Pulleys were invented and engineers and builders were able to raise heavy weights. Since that time, the circle has been the most important geometric shape in the development of all forms of engineering.

Apart from its practical applications, the circle has an aesthetic value which makes it unique among plane figures. The ancients called it 'the perfect curve' and its symmetry and simplicity has led artists and craftspeople to use the circle as a basis for design for many thousands of years.

Definitions

A circle is the locus of a point which moves so that it is always a fixed distance from another stationary point.

Concentric circles are circles that have the same centre.

Eccentric circles are circles that are not concentric.

Figure 3.35 shows some of the parts of the circle.

Constructions

The length of the circumference of a circle is πD or $2\pi R$, where D is the diameter and R the radius of the circle, π is the ratio of the diameter to the circumference and may be taken as 22/7 or, more accurately, as 3.142.

If you need to draw the circumference of a circle (this is required quite often in subsequent chapters), you should either calculate it or use the construction shown in Figure 3.36. This construction is not exact but is accurate enough for most needs. For the sake of thoroughness, the corresponding construction, that of finding the diameter from the circumference, is shown in Figure 3.37.

To construct the circumference of a circle, given the diameter (Figure 3.36).

1   Draw a semi-circle of the given diameter AB, centre O.
2   From B mark off three times the diameter, BC.
3   From O draw a line at 30° to OA to meet the semi-circle in D.
4   From D draw a line perpendicular to OA to meet OA in E.
5   Join EC.

EC is the required circumference.

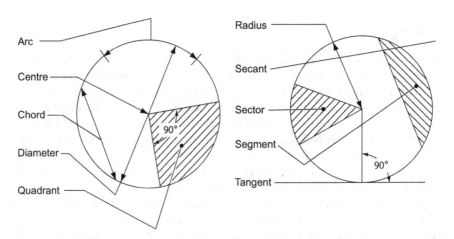

Figure 3.35 Parts of the circle.

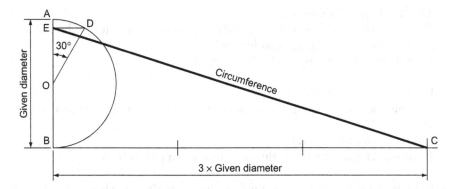

Figure 3.36 To construct the circumference of a circle.

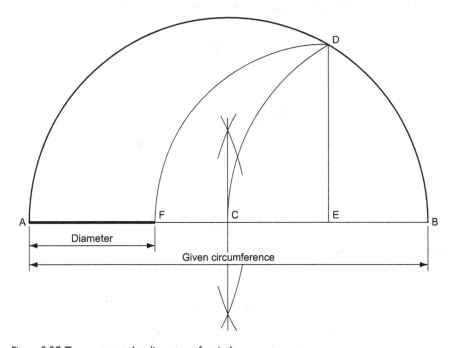

Figure 3.37 To construct the diameter of a circle.

To construct the diameter of a circle, given the circumference (Figure 3.37).

1 Draw the given circumference AB.
2 Bisect AB in C.
3 With centre C and radius CA, draw a semi-circle.
4 With centre B and radius BC, draw an arc to cut the semi-circle in D.
5 From D, draw a perpendicular to AB, to cut AB in E.
6 With centre E and radius ED, draw an arc to cut AB in F.

AF is the required diameter.

The rest of this subsection shows some of the constructions for finding circles drawn to satisfy certain given conditions.

To find the centre of any circle (Figure 3.38).

1 Draw any two chords.
2 Construct perpendicular bisectors to these chords to intersect in O.

O is the centre of the circle.

To construct a circle to pass through three given points (Figure 3.39).

1 Draw straight lines connecting the points as shown. These lines are, in fact, chords of the circle.
2 Draw perpendicular bisectors through these lines to intersect in O.

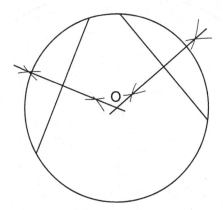

Figure 3.38 To find the centre of any circle.

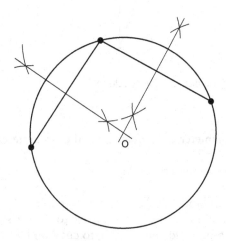

Figure 3.39 To construct a circle to pass through three given points.

O is the centre of a circle that passes through all three points.

To construct the inscribed circle of any regular polygon (in this case, a triangle) (Figure 3.40).

1 Bisect any two of the interior angles to intersect in O. (If the third angle is bisected it should also pass through O.)

O is the centre of the inscribed circle. This centre is called the 'incentre'.

To construct the circumscribed circle of any regular polygon (in this case a triangle) (Figure 3.41).

1 Perpendicularly bisect any two sides to intersect in O. (If the third side is bisected it should also pass through O.)

O is the centre of the circumscribed circle. This centre is called the 'circumcentre'.

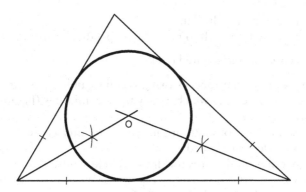

Figure 3.40 To construct the inscribed circle of any regular polygon.

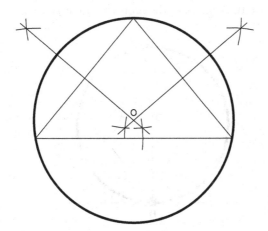

Figure 3.41 To construct the circumscribed circle of any regular polygon.

To construct a circle that passes through a fixed point A and touches a line at a given point B (Figure 3.42).

1   Join AB.
2   From B, erect a perpendicular BC.
3   From A, construct angle BÂO similar to angle CB̂A to intersect the perpendicular in O.

O is the centre of the required circle.

To construct a circle that passes through two given points, A and B, and touches a given line (Figure 3.43).

1   Join AB and produce this line to D (cutting the given line in C) so that BC = CD.
2   Construct a semi-circle on AD.
3   Erect a perpendicular from C to cut the semi-circle in E.
4   Make CF = CE.
5   From F erect a perpendicular.
6   Perpendicularly bisect AB to meet the perpendicular from F in O.

O is the centre of the required circle.

To construct a circle that touches two given lines and passes through a given point P. There are two circles which satisfy these conditions (Figure 3.44).

1   If the two lines do not meet, produce them to intersect in A.
2   Bisect the angle thus formed.
3   From any point on the bisector draw a circle, centre B, to touch the two given lines.
4   Join PA to cut the circle in C and D.
5   Draw PO, parallel to CB and PO, parallel to DB.

O_1 and O_2 are the centres of the required circles.

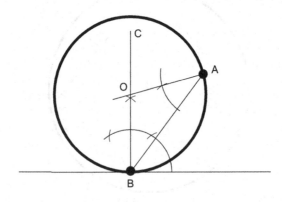

Figure 3.42 To construct a circle that passes through a fixed point.

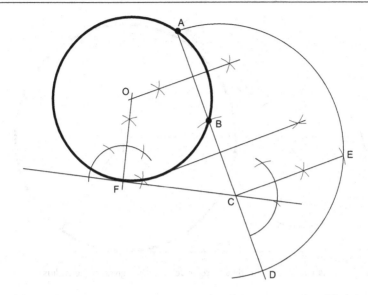

Figure 3.43 To construct a circle that passes through two given points, A and B.

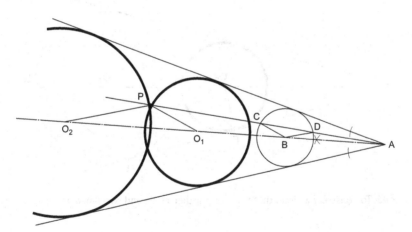

Figure 3.44 To construct a circle that touches two given lines and passes through a given point P.

To construct a circle, radius R, to touch another given circle radius r and a given line (Figure 3.45).

1 Draw a line parallel to the given line, the distance between the lines equal to R.
2 With compass point at the centre of the given circle and radius set at $R + r$, draw an arc to cut the parallel line in O.

O is the centre of the required circle.

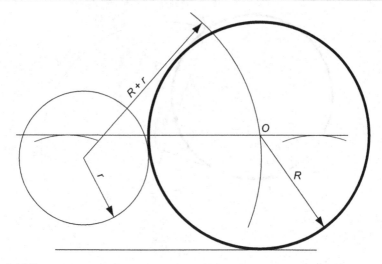

Figure 3.45 To construct a circle, radius R, to touch another given circle radius r.

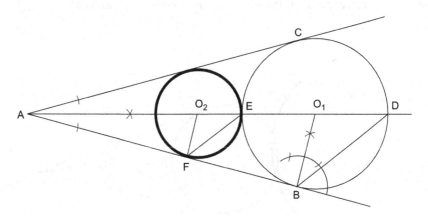

Figure 3.46 To construct a circle that touches another circle and two tangents.

To construct a circle that touches another circle and two tangents of that circle (Figure 3.46).

1   If the tangents do not intersect, produce them to intersect in A.
2   Bisect the angle formed by the tangents.
3   From B, the point of contact of the circle and one of its tangents, construct a perpendicular to cut the bisector in O_1. This is the centre of the given circle.
4   Join BD.
5   Draw EF parallel to DB and FO_2 parallel to BO_1.

   O_2 is the centre of the required circle.

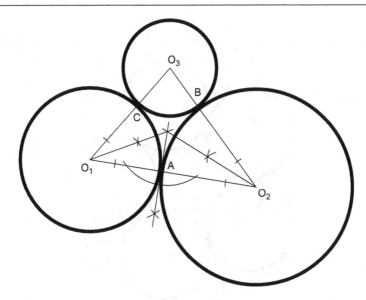

Figure 3.47 To draw three circles which touch each other.

To draw three circles which touch each other, given the position of their cen-tres O_1, O_2 and O_3 (Figure 3.47).

1 Draw straight lines connecting the centres.
2 Find the centre of the triangle thus formed by bisecting two of the interior angles.
3 From this centre, drop a perpendicular to cut O_1O_2 in A.
4 With centre O_1 and radius O_1 A, draw the first circle.
5 With centre O_2 and radius O_2 A, draw the second circle.
6 With centre O_3 and radius O_3 C (= O_3B), draw the third circle.

To draw any number of equal circles within another circle, the circles all to be in contact (in this case 5) (Figure 3.48).

1 Divide the circle into the same number of sectors as there are proposed circles.
2 Bisect all the sectors and produce one of the bisectors to cut the circle in D.
3 From D erect a perpendicular to meet OB produced in E.
4 Bisect DÊO to meet OD in F.
5 F is the centre of the first circle. The other circles have the same radius and have centres on the intersections of the sector bisectors and a circle, centre O and radius OF.

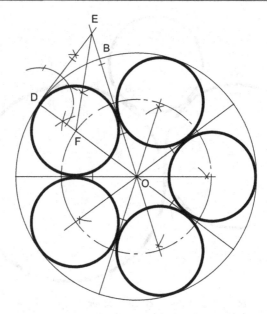

Figure 3.48 To draw any number of equal circles within another circle.

To draw several equal circles within a regular polygon to touch each other and one side of the polygon (in this case, a Heptagon) (Figure 3.49).

1 Find the centre of the polygon by bisecting two of the sides.
2 From this centre, draw lines to all of the corners.
3 This produces several congruent triangles. All we now need to do is to draw the inscribed circle in each of these triangles. This is done by bisecting any two of the interior angles to give the centre C.
4 The circles have equal radii and their centres lie on the intersection of a circle, radius OC and the bisectors of the seven equal angles formed by step 2.

To draw equal circles around a regular polygon to touch each other and one side of the polygon (in this case, a Heptagon) (Figure 3.50).

1 Find the centre of the polygon by bisecting two of the sides.
2 From the centre O draw lines through all of the corners and produce them.
3 Bisect angles CÂB and DB̂A to intersect in E.
4 E is the centre of the first circle. The rest can be obtained by drawing a circle, radius OE, and bisecting the seven angles formed by step 2. The intersections of this circle and these lines give the centres of the other six circles.

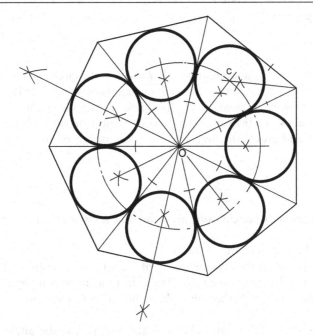

Figure 3.49 To draw several equal circles within a regular polygon.

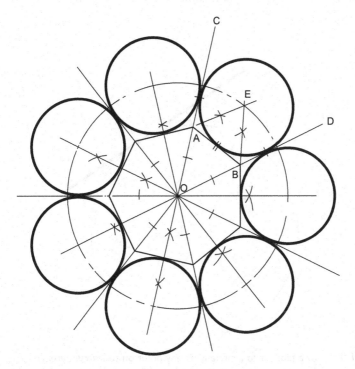

Figure 3.50 To draw equal circles around a regular polygon.

3.2 TANGENCY

Definition

A tangent to a circle is a straight line that touches the circle at one point.

Every curve ever drawn could have tangents drawn to it, but this chapter is concerned only with tangents to circles. These have wide applications in engineering drawing since the outlines of most engineering details are made up of straight lines and arcs. Wherever a straight line meets an arc, a tangent meets a circle.

Constructions

To draw a tangent to a circle from any point on the circumference (Figure 3.51).

1 Draw the radius of the circle.
2 At any point on the circumference of a circle, the tangent and the radius are perpendicular to each other. Thus, the tangent is found by constructing an angle of 90° from the point where the radius crosses the circumference.

A basic geometric theorem (Thales' Theorem) is that the angle in a semicircle is a right angle (Figure 3.52).

This fact is made use of in many tangent constructions.

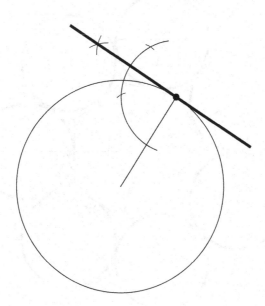

Figure 3.51 To draw a tangent to a circle from any point on the circumference.

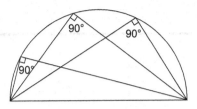

Figure 3.52 Thales' theorem.

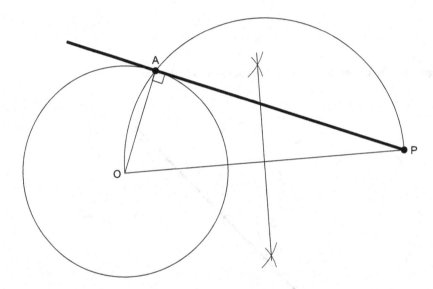

Figure 3.53 To construct a tangent from a point P to a circle.

To construct a tangent from a point P to a circle, centre O (Figure 3.53).

1   Join OP.
2   Erect a semi-circle on OP to cut the circle in A.

PA produced is the required tangent (OA is the radius and is perpendicular to PA since it is the angle in a semi-circle). There are, of course, two tangents to the circle from P but only one has been shown for clarity.

To construct a common tangent to two equal circles (Figure 3.54).

1   Join the centres of the two circles.
2   From each centre, construct lines at 90° to the centre line. The intersection of these perpendiculars with the circles gives the points of tangency.

This tangent is often described as the common exterior tangent.

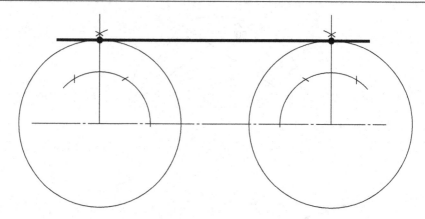

Figure 3.54 To construct a common tangent to two equal circles.

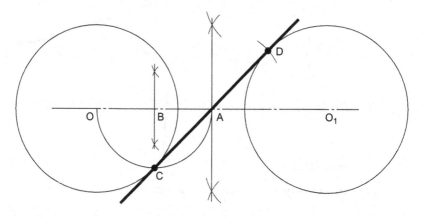

Figure 3.55 To construct the common interior tangent to two equal circles.

To construct the common interior (or transverse or cross) tangent to two equal circles, centres O and O_1 (Figure 3.55).

1   Join the centres OO_1.
2   Bisect OO_1 in A.
3   Bisect OA in B and draw a semi-circle, radius BA to cut the circle in C.
4   With centre A and radius AC, draw an arc to cut the second circle in D.

CO is the required tangent.

To construct the common tangent between two unequal circles, centres O and O_1 and radii R and r, respectively (Figure 3.56).

1   Join the centres OO_1.
2   Bisect OO_1 in A and draw a semi-circle, radius AO.
3   Draw a circle, centre O, radius $R - r$, to cut the semi-circle in B.

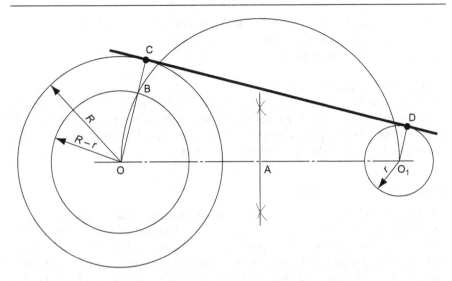

Figure 3.56 To construct the common tangent between two unequal circles.

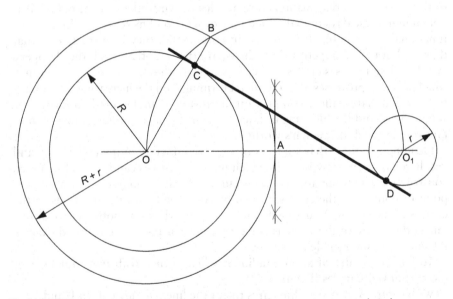

Figure 3.57 To construct the common internal tangent between two unequal circles.

4 Join OB and produce to cut the larger circle in C.
5 Draw O₁D parallel to OC.

CD is the required tangent.
 To construct the common internal tangent between two unequal circles, centres O and O₁ and radii R and r, respectively (Figure 3.57).

1 Join the centres OO₁.
2 Bisect OO₁ in A and draw a semi-circle, radius OA.

3 Draw a circle, centre O, radius $R + r$, to cut the semi-circle in B.
4 Join OB. This cuts the larger circle in C.
5 Draw O_1D parallel to OB.

CD is the required tangent.

A tangent is, by definition, a straight line. However, we do often talk of radii or curves meeting each other tangentially. We mean, of course, that the curves meet smoothly and with no change of shape or bumps. This topic, the blending of lines and curves, is discussed in the following.

Blending of Lines and Curves

It is usually only the very simple type of engineering detail that has an out-line composed entirely of straight lines. The inclusion of curves within the outline of a component may be for several reasons: to eliminate sharp edges, thereby making it safer to handle; to eliminate a stress centre, thereby making it stronger; to avoid extra machining, thereby making it cheaper; and last, but by no means least, to improve its appearance. This last reason applies particularly to those industries that manufacture articles to sell to the general public. It is not enough these days to make vacuum cleaners, food mixers or ballpoint pens functional and reliable. It is equally important that they be attractive so that they, and not the competitors' products, are the ones that catch the shopper's eye. The designer uses circles and curves to smooth out and soften an outline. Machine-shop processes like cold metal forming, and the increasing use of plastics and laminates, allow complex outlines to be manufactured as cheaply as simple ones, and the blending of lines and curves plays an increasingly important role in the draftsperson's world.

Blending is a topic that students often have difficulty in understanding, and yet there are only a few ways in which lines and curves can be blended. When constructing an outline that contains curves blending, do not worry about the point of contact of the curves; rather, be concerned with the positions of the centres of the curves. A curve will not blend properly with another curve or line unless the centre of the curve is correctly found. If the centre is found exactly, the curve is bound to blend exactly.

To find the centre of an arc, radius r, which blends with two straight lines meeting at right angles (Figure 3.58).

With centre A, radius r, draw arcs to cut the lines of the angle in B and C.

With centres B and C, radius r, draw two arcs to intersect in O.

O is the required centre.

This construction applies only if the angle is a right angle. If the lines meet at any angle other than 90°, use the construction shown in Figure 3.59.

To find the centre of an arc, radius r, which blends with two straight lines meeting at any angle (Figure 3.59).

Construct lines, parallel with the lines of the angle and distance r away, to intersect in O.

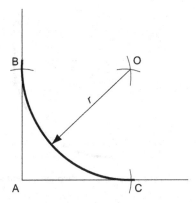

Figure 3.58 To find the centre of an arc which blends with two straight lines at right angles.

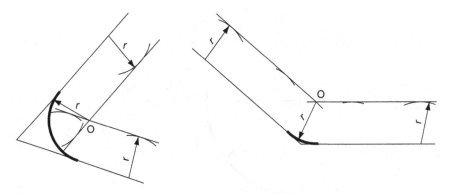

Figure 3.59 To find the centre of an arc which blends with two straight lines at any angle.

O is the required centre.

To find the centre of an arc, radius r, which passes through a point P and blends with a straight line (Figure 3.60).

Construct a line, parallel with the given line, distance r away. The centre must lie somewhere along this line.

With centre P, radius r, draw an arc to cut the parallel line in O.

O is the required centre.

To find the centre of an arc, radius R, which blends with a line and a circle, centre B, radius r.

There are two possible centres, shown in Figure 3.61 and Figure 3.62.

Construct a line, parallel with the given line, distance R away. The centre must lie somewhere along this line.

With centre B, radius $R + r$, draw an arc to intersect the parallel line in O.

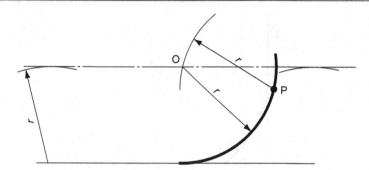

Figure 3.60 To find the centre of an arc which passes through a point and blends with a straight line.

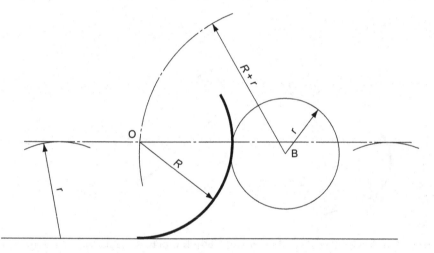

Figure 3.61 To find the centre of an arc which blends with a line and a circle.

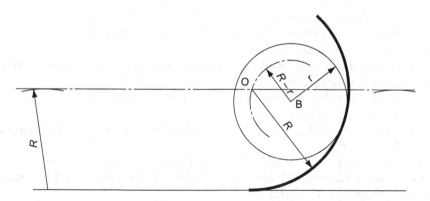

Figure 3.62 Alternative method to find the centre of an arc which blends with a line and a circle.

O is the required centre.

The alternative construction is:

Construct a line, parallel with the given line, distance R away. The centre must lie somewhere along this line.

With centre B, radius $R - r$, draw an arc to intersect the parallel line in O.

O is the required centre.

To find the centre of an arc, radius R, which blends with two circles, centres, respectively.

There are two possible centres, shown in Figure 3.63 and Figure 3.64.

If an arc, radius R, is to blend with a circle, radius r, the centre of the arc must be distance R from the circumference and hence $R + r$ (Figure 3.63) or $R - r$ (Figure 3.64) from the centre of the circle.

With centre A, radius $R + r_1$, draw an arc.

With centre B, radius $R + r_2$, draw an arc to intersect the first arc in O.

O is the required centre.

The alternative construction is:

With centre A, radius $R - r_1$, draw an arc.

With centre B, radius $R - r_2$, draw an arc to intersect the first arc in O.

O is the required centre.

To join two parallel lines with two equal radii, the sum of which equals the distance between the lines (Figure 3.65).

Draw the centre line between the parallel lines.

From a point A, drop a perpendicular to meet the centre line in O_1.

With centre O_1, radius O_1A, draw an arc to meet the centre line in B.

Produce AB to meet the other parallel line in C.

From C erect a perpendicular to meet the centre line in O_2.

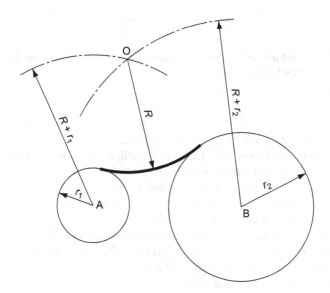

Figure 3.63 To find the centre of an arc which blends with two circles.

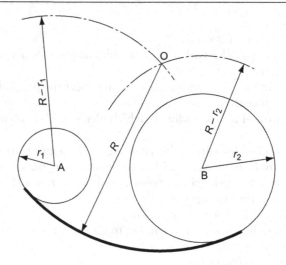

Figure 3.64 Alternative method to find the centre of an arc which blends with two circles.

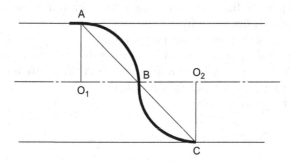

Figure 3.65 To join two parallel lines with two equal radii, the sum of which equals the distance between the lines.

With centre O_2, radius O_2C, draw the arc BC.

To join two parallel lines with two equal radii, r, the sum of which is greater than the distance between the lines (Figure 3.66).

Draw the centre line between the parallel lines.

From a point A, drop a perpendicular and on it mark off $AO_1 = r$.

Draw the centre line between the parallel lines.

From a point A, drop a perpendicular and on it mark off $AO_1 = r$.

With centre O_1, radius r, draw an arc to meet the centre line in B.

Produce AB to meet the other parallel line in C.

From C erect a perpendicular $CO_2 = r$.

With centre O_2, radius r, draw the arc BC.

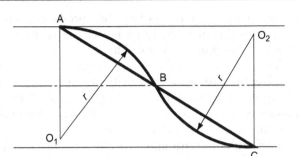

Figure 3.66 To join two parallel lines with two equal radii, the sum of which is greater than the distance between the lines.

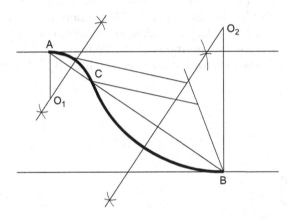

Figure 3.67 To join two parallel lines with two unequal radii given the ends of the curve A and B.

To join two parallel lines with two unequal radii (e.g. in the ratio of 3:1) given the ends of the curve A and B (Figure 3.67).

Join AB and divide into the required ratio, AC:CB = 1:3.

Perpendicularly bisect AC to meet the perpendicular from A in O_1.

With centre O_1, radius O_1A, draw the arc AC.

Perpendicularly bisect CB to meet the perpendicular from B in O_2.

With centre O_2 and radius O_2B, draw the arc CB.

3.3 LOCI

Definition

A locus (plural *loci*) is the path traced out by a point that moves under given definite conditions.

You may not have been aware of it, but you have met loci many times before. One of the most common loci is that of a point that moves so that its distance from another fixed point remains constant: this produces a circle. Another locus that you know is that of a point that moves so that its distance from a line remains constant: this produces parallel lines.

Problems on loci can take several different forms. One important practical application is finding the path traced by points on mechanisms. This may be done simply to see if there is sufficient clearance around a mechanism or, with further knowledge beyond the scope of this book, to determine the velocity and hence the forces acting upon a component.

There are very few rules to learn about loci; it is mainly a subject for common sense. A locus is formed by continuous movement and you have to 'stop' the movement several times and find and plot the position of the point that you are interested in. Take, for instance, the case of the man who was too lazy to put wedges under his ladder. The inevitable happened and the ladder slipped. The path that the feet of the man took is shown in Figure 3.68.

The top of the ladder slips from T to T_9. The motion of the top of the ladder has been stopped at T_1, T_2, T_3, etc., and, since the length of the ladder remains constant, the corresponding positions of the bottom of the ladder, B_1, B_2, B_3, etc., can be found. The positions of the ladder, T_1B_1, T_2B_2, T_3B_3, etc., are drawn

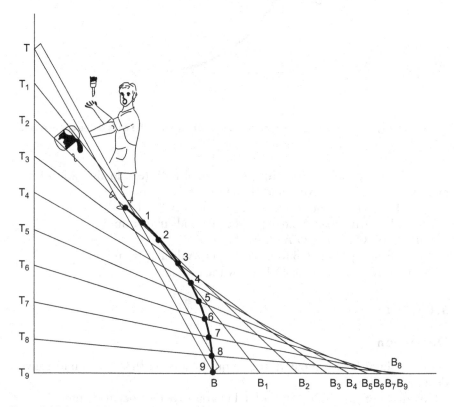

Figure 3.68 Locus of a point on a sliding ladder.

and the position of the man's feet, 1, 2, 3, etc., are marked. The points are joined together with a smooth curve. It is interesting to note that the man hits the ground at right angles (assuming that he remains on the ladder). The resulting jar often causes serious injury and is one of the reasons for using chocks.

3.3.1 Loci of Mechanisms

We now look at some of the loci that can be found on the moving parts of some machines.

Definitions

Velocity is speed in a given direction. It is a term usually reserved for inanimate objects; we talk about the muzzle velocity of a rifle or the escape velocity of a space probe. When we use the word speed we refer only to the rate of motion. When we use the word velocity we refer to the rate of motion and the direction of the motion.

Linear velocity is velocity along a straight line (a linear graph is a straight line).

Angular velocity is movement through a certain angle in a certain time. It makes no allowance for distance travelled. If, as in Figure 3.69, a point P moves through 60° in 1 second, its angular velocity is exactly the same as that of Q, providing that Q also travels through 60° in 1 second. The velocity, as distinct from the angular velocity, will be much greater of course.

Constant velocity, linear or angular, is movement without acceleration or deceleration.

The piston/connecting rod/crank mechanism is very widely used, principally because of its application in internal combustion engines. The piston travels in a straight line; the crank rotates. The connecting rod, which links these two, follows a path that is somewhere in between these two, the exact shape being

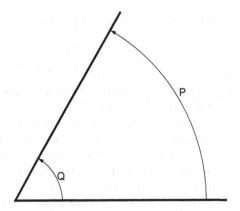

Figure 3.69 Definition of angular velocity.

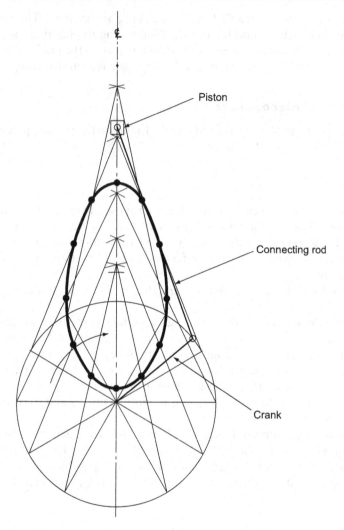

Figure 3.70 Locus of a point halfway up (or down) the rod in a piston/connecting rod/crank mechanism.

dependent on the point of the rod being considered. Figure 3.70 shows the locus of a point halfway up (or down) the rod.

The movement of the crank is continuous. The movement of the piston is also continuous between the top and bottom of its travel. This movement, as before, must be 'stopped' several times and the positions of the centre of the connecting rod found. As with most machines that have cranks, the best policy is to plot the position of the crank in 12 equally spaced positions. This is easily achieved with a 60° set square. The piston must always lie on the centre line and, of course, the connecting rod cannot change its length. It is therefore a simple matter to plot the position of the connecting rod for the 12 positions of the crank. This is best done with compasses or dividers. The midpoint of the connecting rod can then be marked with dividers and the points joined together with a smooth curve.

The direction of rotation of the crank is usually given in problems of this nature. It may make no difference to plotting any of the loci, but it could make a tremendous difference to the functioning of the real machine: what good is a car that does 120 km/h backwards and 10 km/h forwards?

EXAMPLES

Two examples of loci are shown as follows where a point moves relative to another point or to lines.

To plot the locus of a point P that moves so that its distance from a point S and a line XY is always the same (Figure 3.71).

The point S is 20 mm from the line XY.

The first point to plot is the one that lies between S and the line. Since S is 20 mm from the line, and P is equidistant from both, this first point is 10 mm from both.

If we now let the point P be 20 mm from S, it will lie somewhere on the circumference of a circle, centre S, radius 20 mm. Since the point is equidistant from the line, it must also lie on a line drawn parallel to XY and 20 mm away. The second point, then, is the intersection of the 20 mm radius arc and the parallel line.

The third point is at the intersection of an arc, radius 30 mm and centre S, and a line drawn parallel to XY and 30 mm away.

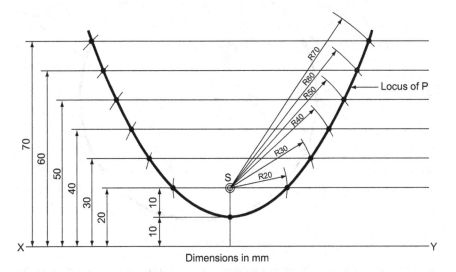

Figure 3.71 Locus of a point that moves equidistant from a given point and a line.

The fourth point is 40 mm from both the line and the point S. This may be continued for as long as is required.

The curve produced is a parabola.

To plot the locus of a point P that moves so that its distance from two fixed points R and S, 50 mm apart, is always in the ratio 2:1, respectively (Figure 3.72).

As in the previous example, the first point to plot is the one that lies between R and S. Since it is twice as far away from R as it is from S, this is done by proportional division of the line RS.

If we now let P be 40 mm from R, it must be 20 mm from S. Thus, the second position of P is at the intersection of an arc, centre R, radius 40 mm and another arc, centre S and radius 20 mm.

The third position of P is the intersection of arcs, radii 50 and 25 mm, centres R and S, respectively.

This is continued for as long as necessary. In this case, at a point 100 mm from R and 50 mm from S, the locus meets itself to form a circle.

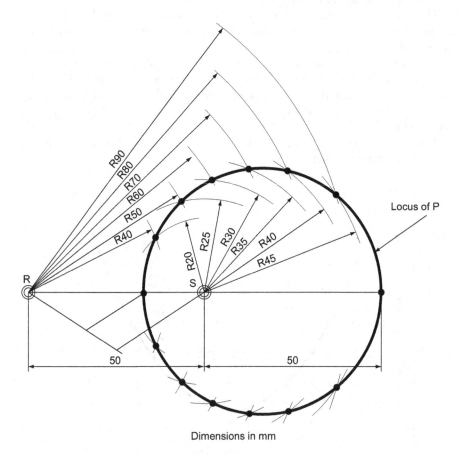

Dimensions in mm

Figure 3.72 Locus of a point that moves so that its distance from two fixed points is at constant ratio.

3.3.2 The Cycloid

The cycloid is the locus of a point on the circumference of a circle as the circle rolls, without slipping, along a straight line.

The approach to plotting a cycloid, as with all problems with loci, is to break down the total movement into a convenient number of parts and consider the conditions at each particular part (Figure 3.73). We have found, when considering the circle, that 12 is the most convenient number of divisions. The total distance that the circle will travel in 1 revolution is πD, the circumference, and this distance is also divided into 12 equal parts. When the circle rolls along the line, the locus of the centre will be a line parallel to the base line and the exact position of the centre will, in turn, be directly above each of the divisions marked off.

If a point P, on the circumference, is now considered, then after the circle has rotated $\dfrac{1}{12}$ of a revolution, point P is somewhere along the line P_1P_{11}. The distance from P to the centre of the circle is still the radius and, thus, if the intersection of the line P_1P_{11} and the radius of the circle marked off from the new position of the centre O_1, is plotted, then this must be the position of the point P after $\dfrac{1}{12}$ of a revolution.

After $\dfrac{1}{6}$ of a revolution the position of P is the intersection of the line P_2P_{10} and the radius, marked off from O_2. This is repeated for the 12 divisions.

Figure 3.73 also shows the beginning of a second cycloid, and it can be seen that the change from one cycloid to another is sudden. If any locus is plotted and has an instantaneous change of shape, it indicates that there is a cessation of movement. Anything that has mass cannot change direction suddenly

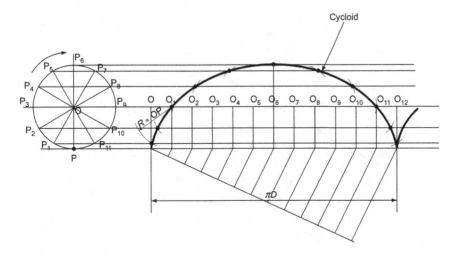

Figure 3.73 The cycloid.

without first ceasing to move. The point of the circle actually in contact with the line is stationary.

This raises the interesting point that, theoretically, a motor car tyre is not moving at all when it is in contact with the road. This is not true in practice, since the contact between the road and tyre is not a point contact, but it does explain why tyres last much longer than would be expected.

At the top of the cycloid, between points 5 and 7, the point P is travelling nearly twice the distance that the centre moves in $\frac{1}{12}$ of a revolution. Thus, a jet car travelling at 800 km/h has points on the rim of the tyre moving up to 1600 km/h – faster than the speed of sound.

The tangent and normal to the cycloid (Figure 3.74).

From the point P, where you wish to draw the normal and the tangent, draw an arc whose radius is the same as the rolling circle, to cut the centre line in O.

With centre O, draw the rolling circle to touch the base line in Q.

PQ is the normal. The tangent is found by erecting a line at 90° to the normal.

The epi-cycloid and the hypo-cycloid (Figure 3.75).

The epi-cycloid is the locus of a point on the circumference of a circle when it rolls, without slipping, along the *outside* of a circular arc.

A hypo-cycloid is the locus of a point on the circumference of a circle when the circle rolls, without slipping, along the *inside* of a circular arc.

The constructions for plotting these curves are very similar to those used for plotting the cycloid.

The circumference of the rolling circle must be plotted along the arc of the base circle. It is possible to calculate this circumference and to plot it along the arc, but this is fairly complicated and it is sufficiently accurate to measure $\frac{1}{12}$ of the circumference of the rolling circle and step this out 12 times, with dividers, along the base arc.

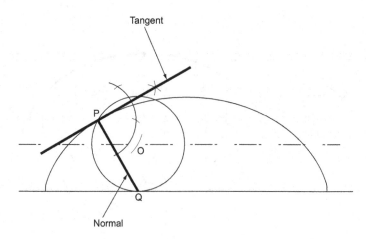

Figure 3.74 The tangent and normal to a cycloid.

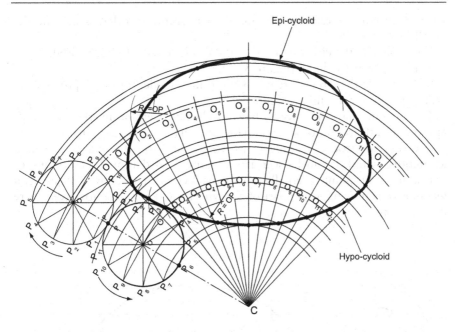

Figure 3.75 The epi-cycloid and the hypo-cycloid.

The remaining construction is similar to that used for the cycloid. The technique is still to plot the intersection of the line drawn parallel to the base, in this case another arc with centre C, and the radius of the rolling circle from its position after $\frac{1}{12}, \frac{1}{6}, \frac{1}{4}$ revolutions, etc.

The main point to watch is that the locus of the centre is no longer coincident with the line P_3P_9 as it was for the cycloid.

The epi-cycloid and the hypo-cycloid form the basis for the shape of some gear teeth, although cycloidal gear teeth have now generally been superseded by gear teeth based on the involute.

The tangent and normal to the epi-cycloid and hypo-cycloid.

The method of obtaining the tangent and normal of an epi-cycloid or a hypo-cycloid is exactly the same as for a cycloid.

3.3.3 The Trochoid

A trochoid is the locus of a point, not on the circumference of a circle but attached to it, when the circle rolls, without slipping, along a straight line.

Again, the technique is similar to that used for plotting the cycloid. The main difference in this case is that the positions of the line P_1P_{11}, P_2P_{10}, etc., are dependent upon the distance of P to the centre O of the rolling circle – not on the radius of the rolling circle as before. This distance PO is also the radius to set on your compasses when plotting the intersections of that radius and the lines P_1P_{11}, P_2P_{10}, etc.

If P is *outside* the circumference of the rolling circle, the curve produced is called a 'superior trochoid' (Figure 3.76).

If P is *inside* the circumference of the rolling circle, the curve produced is called an 'inferior trochoid' (Figure 3.77).

The trochoid has relevance to naval architects. Certain inverted trochoids approximate the profile of waves and therefore have applications in hull design.

The superior trochoid is the locus of the point on the outside rim of a locomotive wheel. It can be seen from Figure 3.76 that at the beginning of a revolution this point is actually moving backwards. Thus, however quickly a locomotive is moving, some part of the wheel is moving back towards where it came from.

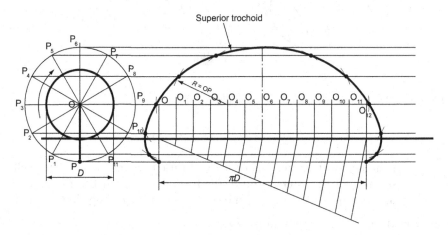

Figure 3.76 The superior trochoid.

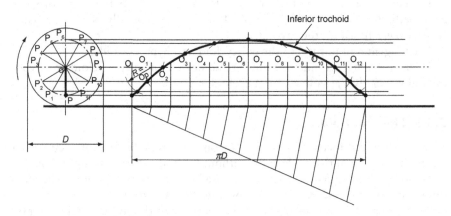

Figure 3.77 The inferior trochoid.

3.3.4 The Involute

There are several definitions for the involute, none being particularly easy to follow.

An involute is the locus of a point, initially on a base circle, which moves so that its straight-line distance, along a tangent to the circle, to the tangential point of contact, is equal to the distance along the arc of the circle from the initial point to the instant point of tangency.

Alternatively, the involute is the locus of a point on a straight line when the straight line rolls round the circumference of a circle without slipping.

The involute is best visualised as the path traced out by the end of a piece of cotton when the cotton is unrolled from its reel.

A quick, but slightly inaccurate, method of plotting an involute is to divide the base circle into 12 parts and draw tangents from the 12 circumferential divisions, Figure 3.78. Measure $\frac{1}{12}$ of the circumference with dividers. When the line has unrolled $\frac{1}{12}$ of the circumference, this distance is stepped out from the tangential point. When the line has unrolled $\frac{1}{6}$ of the circumference, the dividers are stepped out twice. When 1/4 has unrolled, the dividers are stepped out three times, etc. When all 12 points have been plotted, they are joined together with a neat freehand curve.

A more accurate method is to calculate the circumference and lay out this length from a point on the base circle, Figure 3.79. Divide the length into 12 equal parts and use compasses to swing the respective divisions to their inter-sections with the tangents.

The normal and tangent to an involute (Figure 3.80).

The construction for the normal, and hence the tangent, to an involute relies on the construction of a tangent from a point to a circle.

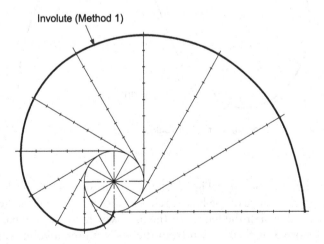

Involute (Method 1)

Figure 3.78 The involute.

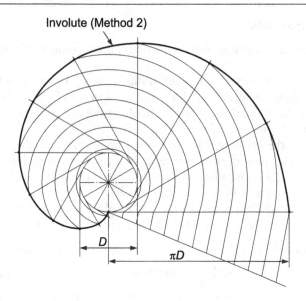

Figure 3.79 Alternative method to construct the involute.

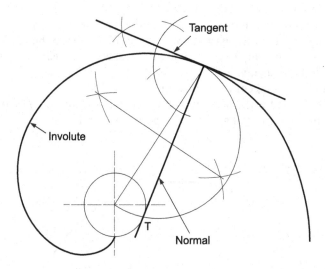

Figure 3.80 The normal and tangent to an involute.

The construction, shown in Figure 3.80, is to draw a line from the point on the involute to the centre of the base circle and bisect it. This gives the centre of a semi-circle, radius half the length of the line, which crosses the base circle at point T. The normal is then drawn from the point on the involute through the point T. The tangent is found by erecting a perpendicular to the normal from T.

3.3.5 The Archimedean Spiral

The Archimedean spiral is the locus of a point that moves away from another fixed point at uniform linear velocity and uniform angular velocity.

It may also be considered to be the locus of a point moving at constant speed along a line when the line rotates about a fixed point at constant speed.

Since both the linear and angular speeds are constant, the only rule for plotting an Archimedean spiral is that the linear and angular distances moved through must both be divided into the same number of equal parts. The most convenient number of equal parts is 12 and if one convolution (when dealing with spirals a movement through 360° is called a convolution as distinct from a revolution) is to be plotted, then the linear distance moved through is divided into 12 equal parts and the 360° into 30° intervals, Figure 3.81. The linear divisions may then be swung round to intersect with the respective angular divisions.

If more than one convolution is to be drawn, then, although the number of angular divisions remains at 12, the linear divisions must be divided into the appropriate multiple. Thus, if two convolutions are to be drawn, there will be 24 linear divisions, etc.

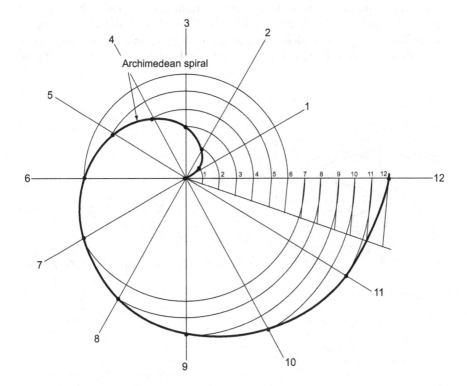

Figure 3.81 The Archimedean spiral.

3.3.6 The Helix

The helix is the locus of a point that moves round a cylinder at constant velocity while advancing along the cylinder at constant velocity.

It is the curve obtained when a piece of string is wound round a cylinder.

It is also the curve generated when turning on a lathe. Normally, the rate of advance along the piece of work is so small that it is indistinguishable as a helix, but it is quite easily seen when cutting a screw thread.

The distance that the point moves along the cylinder in one complete revolution is called 'the pitch'.

The construction of the helix is simple, Figure 3.82. The movement round and along the cylinder is constant and so, for a fixed period, say one complete revolution, the two movements are divided into the same number of equal parts. Thus, $\frac{1}{12}$ of a revolution will coincide with $\frac{1}{12}$ pitch, $\frac{1}{6}$ of a revolution will coincide with $\frac{1}{6}$ pitch, etc.

Figure 3.82 also shows the development of the helix.

Coiled Springs

Most coiled springs are formed on a cylinder and are, therefore, helical. They are, in fact, more often called helical springs than coiled springs. If the spring is to be used in tension, the coils will be close together to allow the spring to stretch. This is the spring that you will see on spring balances in the science lab. If the spring is to be in compression, the coils will be further apart. These

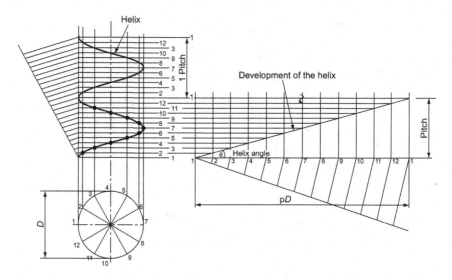

Figure 3.82 The helix.

springs can be seen on the suspension of many modern cars, particularly on the front suspension.

Drawing a helical spring actually consists of drawing two helices, one within another. Although the diameters of the helices differ, their pitch must be the same. Once the points are plotted, it is just a question of sorting out which parts of the helices can be seen and which parts are hidden by the thickness of the wire.

For clarity, the thickness of the wire in Figure 3.83 is 1/4 the pitch of the helix, but if it was not a convenient fraction, it would be necessary to set out the pitch twice. The distance between the two pitches would be the thickness of the wire.

Screw Thread Projection

A screw thread is helical. Unless the screw thread is drawn at a large scale, it is rarely drawn as a helix – except as an exercise in drawing helices!

A good example is to draw a screw thread with a square section. This is exactly the same construction as the coiled spring except that the central core hides much of the construction.

A right-hand screw thread is illustrated in Figure 3.84. To draw a left-hand screw thread merely plot the ascending points from right to left instead of from left to right.

Sometimes a double, triple or even a quadruple start thread is seen, particularly on the caps of some containers where the top needs to be taken off quickly. A multiple start thread is also seen on the starter pinion of motor cars. Multiple start screw threads are used where rapid advancement along a shaft is required. When plotting a double start screw thread, two helices are plotted on the same pitch. The first helix starts at point 1 and the second at point 7. If a triple start screw thread is plotted, the starts are points 1, 9 and 5 (Figure 3.85). If a quadruple start thread is plotted, the starts are points 1, 10, 7 and 4.

3.4 PROBLEMS

(All questions originally set in imperial units.)

Polygons

1 Construct an equilateral triangle with sides 60 mm long.
2 Construct an isosceles triangle that has a perimeter of 135 mm and an altitude of 55 mm.
3 Construct a triangle with base angles 60° and 45° and an altitude of 76 mm.
4 Construct a triangle with a base of 55 mm, an altitude of 62 mm and a vertical angle of 37½°.
5 Construct a triangle with a perimeter measuring 160 mm and sides in the ratio 3:5:6.

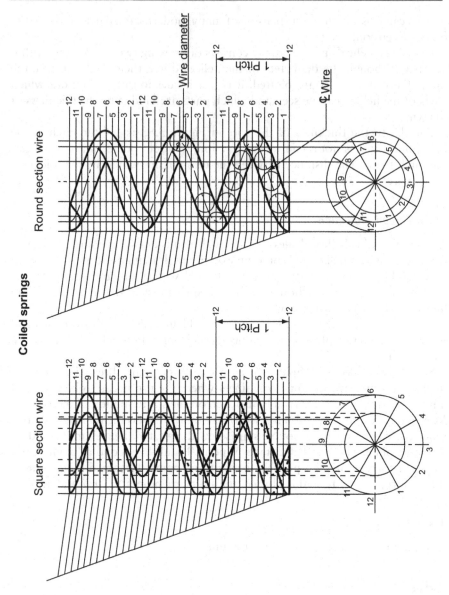

Figure 3.83 Coiled springs.

6   Construct a triangle with a perimeter of 170 mm and sides in the ratio 7:3:5.

7   Construct a triangle given that the perimeter is 115 mm, the altitude is 40 mm and the vertical angle is 45°.

8   Construct a triangle with a base measuring 62 mm, an altitude of 50 mm and a vertical angle of 60°. Now draw a similar triangle with a perimeter of 250 mm.

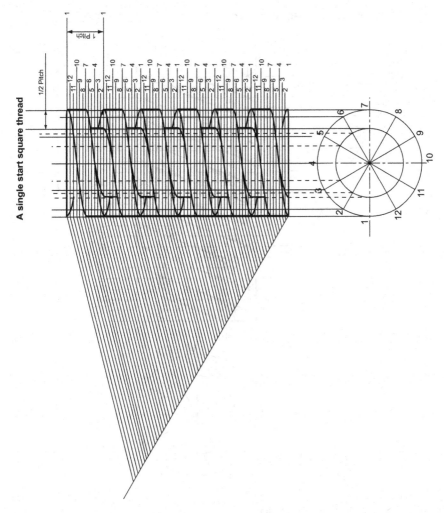

Figure 3.84 A right-hand screw thread.

9 Construct a triangle with a perimeter of 125 mm, whose sides are in the ratio 2:4:5. Now draw a similar triangle whose perimeter is 170 mm.

10 Construct a square of side 50 mm. Find the midpoint of each side by construction and join up the points with straight lines to produce a second square.

11 Construct a square whose diagonal is 68 mm.

12 Construct a square whose diagonal is 85 mm.

13 Construct a parallelogram given two sides 42 mm and 90 mm long, and the angle between them 67°.

14 Construct a rectangle that has a diagonal 55 mm long and one side 35 mm long.

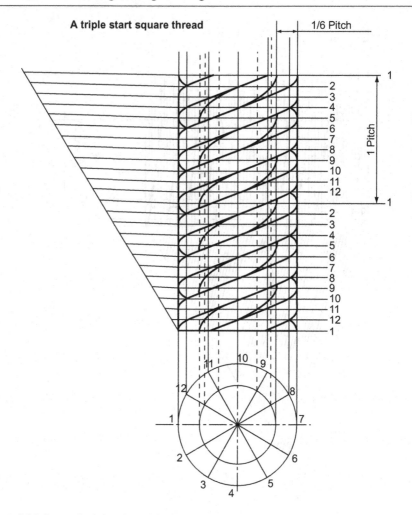

Figure 3.85 Screw thread with triple start.

15 Construct a rhombus if the diagonal is 75 mm long and one side is 44 mm long.

16 Construct a trapezium given that the parallel sides are 50 mm and 80 mm long and are 45 mm apart.

17 Construct a regular hexagon, 45 mm side.

18 Construct a regular hexagon if the diameter is 75 mm.

19 Construct a regular hexagon within an 80 mm diameter circle. The corners of the hexagon must all lie on the circumference of the circle.

20 Construct a square, side 100 mm. Within the square, construct a regular octagon. Four alternate sides of the octagon must lie on the sides of the square.

21 Construct the following regular polygons:

a pentagon, side 65 mm;
a heptagon, side 55 mm;
a nonagon, side 45 mm;
a decagon, side 35 mm.

22 Construct a regular pentagon, diameter 82 mm.

23 Construct a regular heptagon within a circle, radius 60 mm. The corners of the heptagon must lie on the circumference of the circle.

Circles

24 Construct a regular octagon on a base line 25 mm long and draw the inscribed circle. Measure and state the diameter of this circle in mm.

North Western Secondary School Examinations Board

25 Describe *three* circles, each one touching the other two externally, their radii being 12, 18 and 24 mm, respectively.

North Western Secondary School Examinations Board

26 No construction has been shown in Figure 3.86. You are required to draw the figure full size showing all construction lines necessary to ensure the circles are tangential to their adjacent lines.

Southern Regional Examinations Board

27 Construct the triangle ABC in which the base BC = 108 mm, the vertical angle Â = 70° and the altitude is 65 mm.

D is a point on AB 34 mm from A. Describe a circle to pass through the points A and D and touch (tangential to) the line BC.
Southern Universities' Joint Board

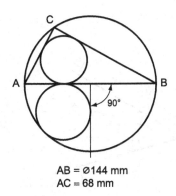

AB = ⌀144 mm
AC = 68 mm

Figure 3.86

28 Figure 3.87 shows two touching circles placed in the corner made by two lines which are perpendicular to one another. Draw the view shown and state the diameter of the smaller circle. Your construction must show clearly the method of obtaining the centre of the smaller circle.

University of London School Examinations

29 Figure 3.88 shows two intersecting lines AB and BC and the position of a point P. Draw the given figure and find the centre of a circle that will pass through P and touch the lines AB and BC. Draw the circle and state its radius as accurately as possible.

University of London School Examinations

30 A triangle has sides 100, 106 and 60 mm long. Draw the triangle and construct the following: (a) the inscribed circle, (b) the circumscribed circle, (c) the smallest escribed circle.

University of London School Examinations

31 Construct an isosceles triangle ABC where the included angle A = 67½°, and AB = AC = 104 mm. Draw circles of 43, 37 and 32 mm radius using as centres A, B and C, respectively.

Construct the smallest circle that touches all three circles.

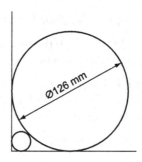

Figure 3.87

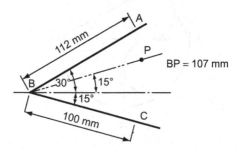

Figure 3.88

Measure and state the diameter of the constructed circle.
Associated Examining Board

32 AB and AC are two straight lines which intersect at an angle of 30°. D is a point between the two lines at perpendicular distances of 37 and 62 mm, respectively from AB and AC. Describe the circle that touches the two converging lines and passes through point D; the centre of this circle is to lie between the points A and D. Now draw two other circles each touching the constructed circle externally and also the converging lines. Measure and state the diameters of the constructed circles.

Oxford Local Examinations

33 OA and OB are two straight lines meeting at an angle of 30°. Construct a circle of diameter 76 mm to touch these two lines and a smaller circle that will touch the two converging lines and the first circle.

Also construct a third circle of diameter 64 mm which touches each of the other two circles.
Oxford Local Examinations

34 Construct a regular octagon of side 75 mm and within this octagon describe eight equal circles each touching one side of the octagon and two adjacent circles. Now draw the smallest circle that will touch all eight circles. Measure and state the diameter of this circle.

Oxford Local Examinations

Tangents

35 A former in a jig for bending metal is shown in Figure 3.89.

(a) Draw the former, full size, showing in full the construction for obtaining the tangent joining the two arcs.
(b) Determine, without calculation, the centres of the four equally spaced holes to be bored in the positions indicated in the figure.
Middlesex Regional Examining Board

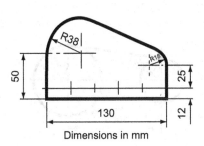

Dimensions in mm

Figure 3.89

36 Figure 3.90 shows a centre finder, or centre square in position on a 75 mm diameter bar.

Draw, full size, the shape of the centre finder and the piece of round bar. Show clearly the constructions for:

(a) the tangent, AA, to the two arcs;
(b) the points of contact and the centre for the 44 mm radius at B;
(c) the points of contact and the centre for the 50 mm radius at C.

South-East Regional Examinations Board

37 Figure 3.91 shows the outline of two pulley wheels connected by a belt of negligible thickness. To a scale of $\dfrac{1}{10}$ draw the figure showing the construction necessary to obtain the points of contact of the belt and pulleys.

Middlesex Regional Examining Board

38

(a) Draw the figure ABCP shown in Figure 3.92 and construct a circle, centre O, to pass through the points A, B and C.
(b) Construct a tangent to this circle touching the circle at point B.

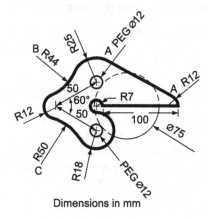

Dimensions in mm

Figure 3.90

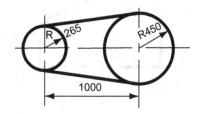

Figure 3.91

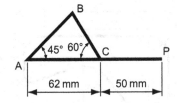

Figure 3.92

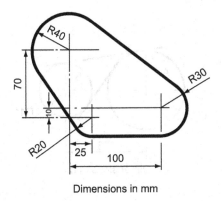

Dimensions in mm

Figure 3.93

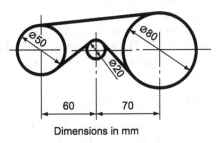

Dimensions in mm

Figure 3.94

(c) Construct a tangent from the point P to touch the circle on the minor arc of the chord AC.

Southern Regional Examining Board

39 Figure 3.93 shows a metal blank. Draw the blank, full size, showing clearly the constructions for obtaining the tangents joining the arcs.

40 Figure 3.94 shows the outlines of three pulley wheels connected by a taut belt. Draw the figure, full size, showing clearly the constructions for obtaining the points of contact of the belt and pulleys.

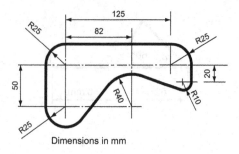

Figure 3.95

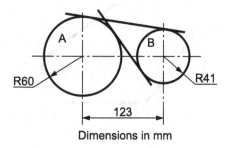

Figure 3.96

41 Figure 3.95 shows the outline of a metal blank. Draw the blank, full size, showing clearly the constructions for finding exact positions of the tangents joining the arcs.

42 A segment of a circle stands on a chord AB which measures 50 mm. The angle in the segment is 55°. Draw the segment. Produce the chord AB to C making BC 56 mm long. From C construct a tangent to the arc of the segment.

University of London School Examinations

43 A and B are two points 100 mm apart. With B as centre draw a circle 75 mm diameter. From A draw two lines AC and AD which are tangential to the circle AC 150 mm. From C construct another tangent to the circle to form a triangle ACD. Measure and state the lengths CD and AD, also angle CDA.

Joint Matriculation Board

44 Figure 3.96 shows two circles, A and B, and a common external tangent and a common internal tangent. Construct (a) the given circles and tangents and (b) the smaller circle that is tangential to circle B and the two given tangents.

Measure and state the distance between the centres of the constructed circle and circle A.

Associated Examining Board

45 Figure 3.97 shows an exhaust pipe gasket. Draw the given view full size and show any constructions used in making your drawing. Do not dimension your drawing.

Southern Regional Examinations Board

46 Figure 3.98 is an elevation of the turning handle of a can opener. Draw this view, twice full size, showing clearly the method of establishing the centres of the arcs.

East Anglian Examinations Board

47 Figure 3.99 shows the outline of an electric lamp.

Important – Construction lines must be visible, showing clearly how you obtained the centres of the arcs and the exact positions of the junctions between arcs and straight lines.
Part 1. Draw the shape, full size.
Part 2. Line AB is to be increased to 28 mm. Construct a scale and using this scale draw the left half or right half of the shape, increasing all other dimensions proportionally.

Southern Regional Examinations Board

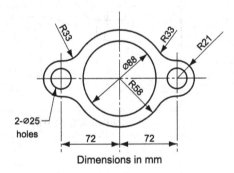

Figure 3.97

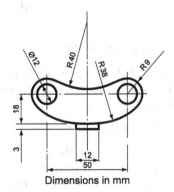

Figure 3.98

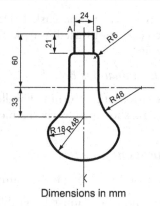

Dimensions in mm

Figure 3.99

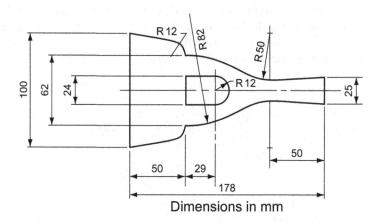

Dimensions in mm

Figure 3.100

48   Figure 3.100 shows a garden hoe. Draw this given view, full size, and show any construction lines used in making the drawing. Do not dimension the drawing.

Southern Regional Examinations Board

49   Figure 3.101 shows one-half of a pair of pliers. Draw, full size, a front elevation looking from A. Your constructions for finding the centres of the arcs must be shown.

South-East Regional Examinations Board

50   Figure 3.102 shows the design for the profile of a sea wall. Draw the profile of the sea wall to a scale of 10 mm = 2 m. Measure in metres the dimensions A, B, C and D and insert these on your drawing. In order to

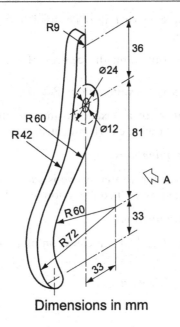

Dimensions in mm

Figure 3.101

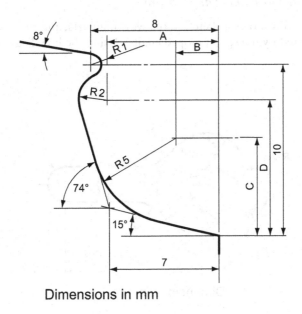

Dimensions in mm

Figure 3.102

do this you should construct an open divided scale of 10 mm = 2 m to show units of 1 m.

Constructions for obtaining the centres of the radii must be clearly shown.

Metropolitan Regional Examinations Board

51 Details of a spanner for a hexagonal nut are shown in Figure 3.103. Draw this outline showing clearly all constructions. Scale: full size.

Oxford Local Examinations

52 The end of the lever for a safety valve is shown in Figure 3.104. Draw this view, showing clearly all construction lines. Scale: ½ full size.

Oxford Local Examinations

53 Draw, to a scale of 2:1, the front elevation of a rocker arm as illustrated in Figure 3.105.

Oxford Local Examinations

Loci

54 Figure 3.106 shows a door stay as used on a wardrobe door. The door is shown in the fully open position. Draw, full size, the locus of end A of the stay as the door closes to the fully closed position. The stay need only be shown diagrammatically as in Figure 3.107.

West Midlands Examinations Board

55 Figure 3.108 shows a sketch of the working parts, and the working parts represented by lines, of a moped engine. Using the line diagram only, and

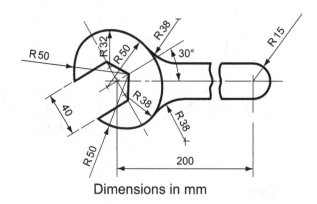

Dimensions in mm

Figure 3.103

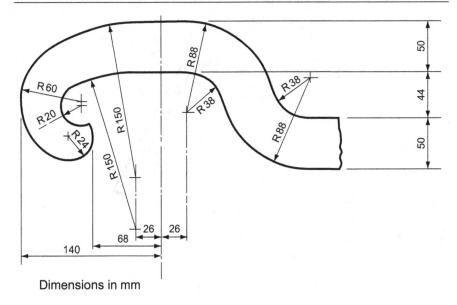

Dimensions in mm

Figure 3.104

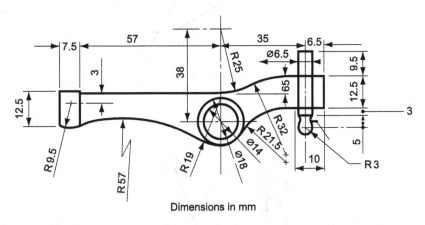

Dimensions in mm

Figure 3.105

drawing in single lines only, plot, full size, the locus of the point P for one full turn of the crank BC.

Do not attempt to draw the detail shown in the sketch. Show all construction.

East Anglian Examinations Board

56 In Figure 3.109 the crank C rotates in a clockwise direction. The rod PB is connected to the crank at B and slides through the pivot D.

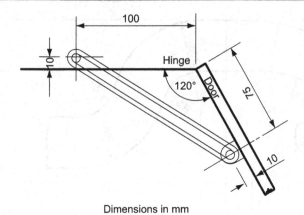

Dimensions in mm

Figure 3.106

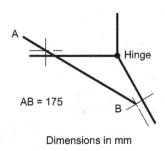

AB = 175

Dimensions in mm

Figure 3.107

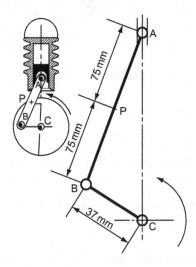

Figure 3.108

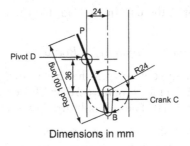

Dimensions in mm

Figure 3.109

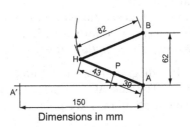

Dimensions in mm

Figure 3.110

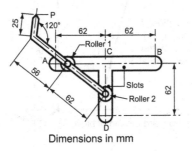

Dimensions in mm

Figure 3.111

Plot, to a scale 1½ full size, the locus of P for one revolution of the crank.

South-East Regional Examinations Board

57 In Figure 3.110 the stay BHA is pinpointed at H and is free to rotate about the fixed point B. Plot the locus of P as end A moves from A to A'.

North Western Secondary School Examinations Board

58 In Figure 3.111, rollers 1 and 2 are attached to the angled rod. Roller 1 slides along slot AB while roller 2 slides along CD and back. Draw, full

size, the locus of P, the end of the rod, for the complete movement of roller 1 from A to B.

South-East Regional Examinations Board

59 As an experiment, a very low gear has been fitted to a bicycle. This gear allows the bicycle to move forward 50 mm for every 15 degrees rotation of the crank and pedal. These details are shown in Figure 3.112.

  (a) Draw, half-full size, the crank and pedal in position as it rotates for every 50 mm forward motion of the bicycle up to a distance of 600 mm. The first forward position has been shown on the drawing.
  (b) Draw a smooth freehand curve through the positions of the pedal which you have plotted.
  (c) From your drawing find the angle of the crank OA to the horizontal when the bicycle has moved forward 255 mm.
  Metropolitan Regional Examinations Board

60 Figure 3.113 is a line diagram of a slotted link and crank of a shaping machine mechanism. The link AC is attached to a fixed point A about which it is free to move about the fixed point on the disc. The disc rotates

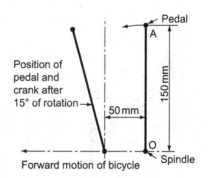

Figure 3.112

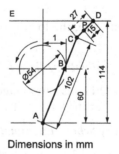

Dimensions in mm

Figure 3.113

about centre O. Attached to C and free to move easily about the points C and D is the link CD. D is also free to slide along DE.

When the disc rotates in the direction of the arrow, plot the locus of C, the locus of P on the link CD, and clearly show the full travel of B on AC.

Southern Universities' Joint Board

61 In Figure 3.114, MP and NP are rods hinged at P, and A and D are guides through which MP and NP are allowed to move. D is allowed to move along BC, but rod NP is always perpendicular to BC. The guide A is allowed to rotate about its fixed point. Draw the locus of P above AB for all positions and when P is always equidistant from A and BC.

This locus is part of a recognised curve. Name the curve and the parts used in its construction.

Southern Universities' Joint Board

62 In the mechanism shown in Figure 3.115, OA rotates about O, PC is pivoted at P and QB is pivoted at Q. BCDE is a rigid link. OA= PC = CD = DE = 25 mm, BC = 37.5 mm, QB = 50 mm and AD = 75 mm. Plot the complete locus of E.

Oxford and Cambridge Schools Examinations Board

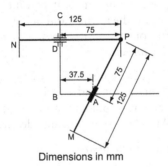

Dimensions in mm

Figure 3.114

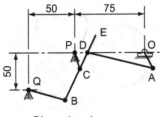

Dimensions in mm

Figure 3.115

63   A rod AB 70 mm long rotates at a uniform rate about end A. Plot the path of a slider S, initially coincident with A, which slides along the rod, at a uniform rate, from A to B and back to A during one complete revolution of the rod.

Joint Matriculation Board

64   With a permanent base of 100 mm, draw the locus of the vertices of all the triangles with a constant perimeter of 225 mm.

Oxford and Cambridge Schools Examination Board

65   Three circles lie in a plane in the positions shown in Figure 3.116. Draw the given figure and plot the locus of a point which moves so that it is always equidistant from the circumferences of circles A and B. Plot also the locus of a point which moves in like manner between circles A and C.

Finally draw a circle whose circumference touches the circles A, B and C and measure and state its diameter.
Cambridge Local Examinations Board

66   Figure 3.117 shows a circular wheel 50 mm in diameter with a point P attached to its periphery. The wheel rolls without slipping along a perfectly straight track whilst remaining in the same plane.

Plot the path of point P for one-half revolution of the wheel on the track. Construct also the normal and tangent to the curve at the position reached after one-third of a revolution of the wheel.
Cambridge Local Examinations

67   The views in Figure 3.118 represent two discs which roll along AB. Both discs start at the same point and roll in the same direction. Plot the curves for the movement of points p and q and state the perpendicular height of p above AB where q again coincides with line AB.

Southern Universities' Joint Board

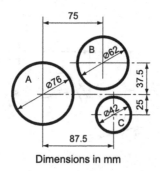

Dimensions in mm

Figure 3.116

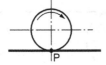

Figure 3.117

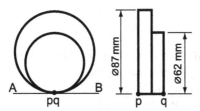

Figure 3.118

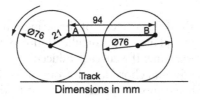

Track
Dimensions in mm

Figure 3.119

68 A wheel of 62 mm diameter rolls without slipping along a straight path. Plot the locus of a point P on the rim of the wheel and initially in contact with the path, for one-half revolution of the wheel along the path. Also construct the tangent, normal and centre of curvature at the position reached by the point P after one quarter revolution of the wheel along the path.

Cambridge Local Examinations

69 The driving wheels and coupling rod of a locomotive are shown to a reduced scale in Figure 3.119. Draw the locus of any point P on the link AB for one revolution of the driving wheels along the track.

University of London School Examinations

70 A piece of string AB, shown in Figure 3.120, is wrapped around the cylinder, centre O, in a clockwise direction. The length of the string is equal to the circumference of the cylinder.

(a) Show, by calculation, the length of the string, correct to the nearest 1 mm, taking π as 3.14.

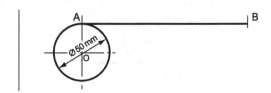

Figure 3.120

(b) Plot the path of the end B of the string as it is wrapped round the cylinder, keeping the string taut.

(c) Name the curve you have drawn.

Middlesex Regional Examining Board

71 A cylinder is 48 mm diameter and a piece of string is equal in length to the circumference. One end of the string is attached to a point on the cylinder.

(a) Draw the path of the free end of the string when it is wound round the cylinder in a plane perpendicular to the axis of the cylinder.

(b) In block letters, name the curve produced.

(c) From a point 56 mm chord length from the end of the curve (i.e. the free end of the string), construct a tangent to the circle representing the cylinder.

Southern Universities' Joint Board

72 A circle 50 mm diameter rests on a horizontal line. Construct the involute to this circle, making the last point on the curve 2π mm from the point at which the circle makes contact with the horizontal line.

Cambridge Local Examinations

73 P, O and Q are three points in that order on a straight line so that PO 34 mm and OQ 21 mm. O is the pole of an Archimedean spiral. Q is the nearest point on the curve and P another point on the first convolution of the curve. Draw the Archimedean spiral showing two convolutions.

Southern Universities' Joint Board

74 Draw two convolutions of an Archimedean spiral such that in two revolutions the radius increases from 18 to 76 mm.

Oxford Local Examinations

75 A piece of cotton is wrapped around the cylinder shown in Figure 3.121. The cotton starts at C and after one turn passes through D, forming a

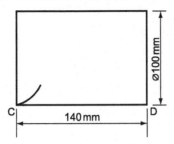

Figure 3.121

helix. The start of the helix is shown in the figure. Construct the helix, showing hidden detail.

Middlesex Regional Examining Board

76 A cylinder, made of transparent material, 88 mm O/D, 50 mm I/D and 126 mm long, has its axis parallel to the VP. Two helical lines marked on its curved surface – one on the outside and the other on the inside – have a common pitch of 63 mm.

Draw the elevation of the cylinder, showing both helices starting from the same radial line and completing two turns.

Associated Examining Board

77 Draw a longitudinal elevation, accurately projected, showing two turns of a helical spring. The spring is of 100 mm outside diameter, the pitch of the coils is 62 mm and the spring material is of 10 mm diameter.

Cambridge Local Examinations

Chapter 4

Methods of Spatial Visualisation

4.1 INTRODUCTION

Graphic representations have always been a basic, natural form of communication. While during the first centuries those language-independent graphics were considered to be artistic drawings, more and more technical drawings emerged. Especially medieval painters significantly influenced the development of rather technical representations by using descriptive geometry to describe three-dimensional objects by two-dimensional drawings. The objective was to represent objects in a spatial view to create a visual impact on the viewer. While details of an object such as dimensions and geometrical features could be communicated to others more comprehensibly, the effort to create such drawings was immense. With the further development of industrialisation, more and more technical representations were needed to create awareness of the object's form and shape which was progressively realised by one-view or multi-view projections.

To represent geometric objects through a two-dimensional image, we need to consider instructions that are fundamental topics of descriptive geometry. Apart from the geometric constructions we learned in Chapter 3, we need to understand the different mapping rules and important characteristics of the geometric entities.

To receive images from a spatial object on a plane surface, descriptive geometry uses the method of *projection*.

Definition

Projections are mappings of objects, i.e. a set of points, on a (usually) plane surface by means of projection lines, the *projectors*. The plane on which the object is mapped is known as the *plane of projection*.

When we look at an object in space, the visible part of it is projected on our eye's retina, the plane of projection. The type of projection that creates the mapped object in our eye is referred to as *central projection*. By using this projection method, the projection centre, also known as *station point* (SP), is placed at a finite distance apart from the projection plane, and all projectors are converging. This kind of projection is called *perspective projection* and is used to pictorially represent a 3D object on 2D paper or a 2D computer screen with just one single view. The reason this representation is sometimes used is that it mimics the human eye. While this might be suitable for providing an impression of three dimensions, this kind of projection is rarely used in mechanical engineering.

In perspective projection, any parallel lines not parallel to the projection plane converge at the same single point, the *vanishing point*. This phenomenon can be observed on any photo taken from a building or from an object which consists of parallel faces and edges. The closer we get to the projection plane, the more the edges become oblique.

Depending on whether the plane of projection is between the observer and the object or behind the object, the projection appears smaller (scaled down) or larger (scaled up) than the original (Figure 4.1).

When we start to increase the distance between the station point and the projection plane, the edges become more parallel. At infinity, they will be finally exactly parallel to each other. This scenario is called *parallel projection*. The housing of a control cabinet, shown in Figure 4.2 as a perspective projection and as a parallel projection, illustrates this effect.

Characteristic of parallel projection is that parallel lines projected on the plane of projection remain parallel. Furthermore, length ratios and division ratios also remain constant. Another advantage of parallel projection is that it does not produce any perspective foreshortening. Figure 4.3 depicts a parallel projection of a triangular face. While the plane of projection P_1 is perpendicular to the projection lines s_n, projection plane P_2 is at an angle to the projectors. Both projected triangles will show the same length ratios.

To classify the types of projection, we need to distinguish whether the projectors are parallel to each other. Another criterion for classification is the relative position of the object to be projected with respect to the plane of projection.

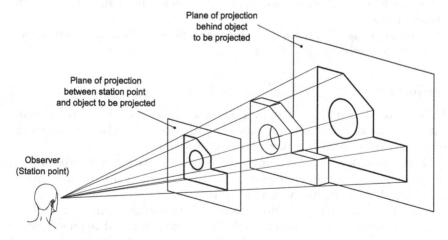

Figure 4.1 Scaling in perspective projection.

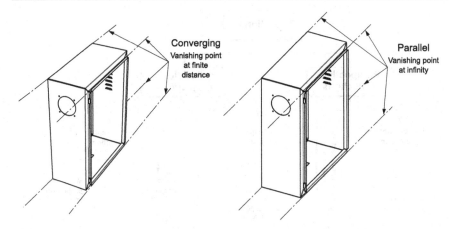

Figure 4.2 Control cabinet housing shown in parallel projection and in perspective projection.

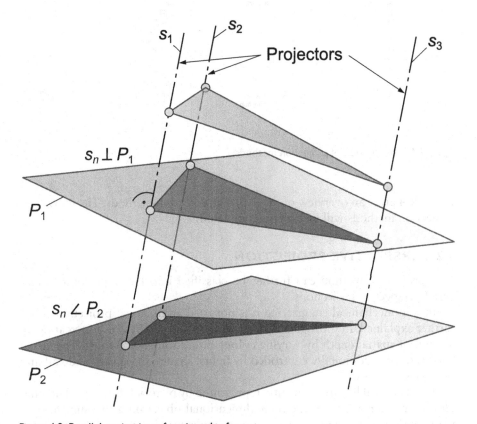

Figure 4.3 Parallel projection of a triangular face.

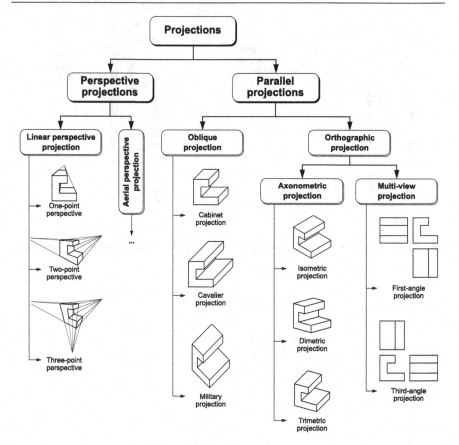

Figure 4.4 Classification of projection methods.

Figure 4.4 gives an overview of the existing projection methods. The individual projection methods will be explained in detail.

4.2 PERSPECTIVE PROJECTION

This projection method can further be classified into linear perspective and aerial perspective projections.

Since in mechanical engineering it is not used at all, aerial projection is not further explained. In short, it is a projection method which aims for creating an illusion of spatial depth by varying colour saturation and contrast. The visual perspective is additionally controlled by lights, shades, the clarity of detail and the levels of tones.

Objects in real life are frequently represented by pictorial drawings. This enables the representation of the three-dimensional object on a two-dimensional plane with a single view. One method to implement is linear perspective projection. Depending on the number of vanishing points (VP), linear perspective

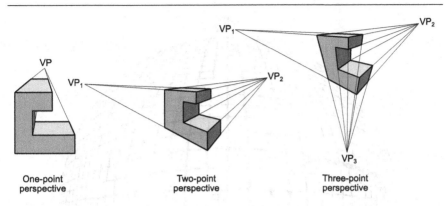

Figure 4.5 Perspective projection methods with one or more vanishing points.

projections can be distinguished between one-point perspective, two-point perspective and three-point perspective (Figure 4.5).

One-point perspective describes a perspective representation of an object placed with one of its faces parallel to the plane of projection. That means that all lines that are neither horizontal nor vertical converge to a common point in the distance.

A one-point perspective, seen from beneath, on a horizontal plane is referred to as a 'frog's eye perspective' or an 'ant's eye perspective', whereas if seen from above it is called a 'bird's eye perspective'.

Two-point perspective projections have two vanishing points, i.e. all horizontal lines or all vertical lines are parallel to each other, but not both at the same time.

In three-point perspective, all lines converge to three different vanishing points and therefore no lines are parallel.

Theoretically, even four-point and five-point perspective exists. However, these projection methods are never used in engineering. Four-point perspective differs from the aforementioned projection methods since it uses curved perspective lines instead of straight lines. In five-point perspective, four vanishing points are equally distributed on a circle while a fifth vanishing point is located at the centre of the circle (Figure 4.6). The resulting perspective is sometimes also called 'fisheye perspective'.

Linear perspective projection is useful to pictorially represent an object by giving the impression of spatial depth. However, closer features appear larger than they are in reality. Apart from one-point perspective, this projection method does not provide the true size of any geometric entity. This is the reason why perspective projection is not used in engineering drawings created for manufacturing. However, this projection method might be suitable for conveying an idea or rough concept.

Figure 4.7 illustrates perspective projection.

For further details regarding basic rules for the development and application of central projection, refer to ISO 5456–4.

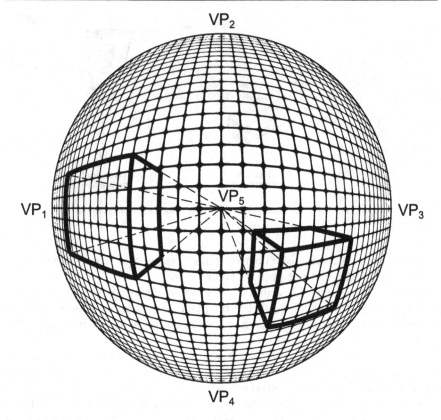

Figure 4.6 Five-point perspective.

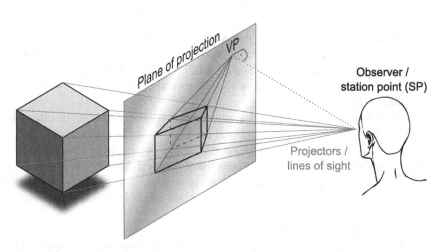

Figure 4.7 Perspective projection.

4.3 OBLIQUE PROJECTION

Oblique parallel projection is another method of pictorial drawing. It is simpler than perspective or isometric projection, which will be introduced in the next section, but it does not present so realistic a picture. In oblique projection the direction vectors of the projection plane's normal and the line of sight are different. In contrast, in orthographic projection they are identical. The various types of parallel projection differ in the angles between the projection plane's normal and the direction of projection as well as between the projectors and the coordinate axes. Figure 4.8 shows the principle of an oblique projection in which all projectors intersect the plane of projection at the same angle other than 0° or 90°.

Figure 4.9 shows a shaped block drawn three times in oblique projection but at different angles. All drawings show the front face of the block drawn in the plane of the paper and the side and top faces receding at 30°, 45° and 60° respectively. An oblique line is one which is neither vertical nor horizontal, and the receding lines in oblique projection can be at any angle other than 0° or 90° as long as they remain parallel in any one drawing. In practice, for manual drafting it is usual to keep to the set square angles and, of the three to choose from, 45° is the most widely used.

If you check the measurements on the oblique drawings with those on the isometric sketch given in the lower right corner of Figure 4.9, you will find that the measurements on the front and oblique faces are all true lengths. This gives rise to a distorted effect. The drawings of the block in the oblique view appear to be out of proportion, particularly when compared with the isometric view.

Figure 4.10 shows how we attempt to overcome this distortion. The oblique lengths have been altered. The degree of alteration has been determined by the oblique angle. An oblique angle of 60° causes a large distortion and the oblique

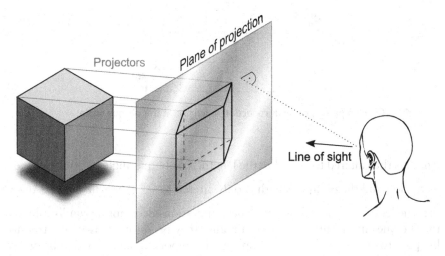

Figure 4.8 Oblique projection.

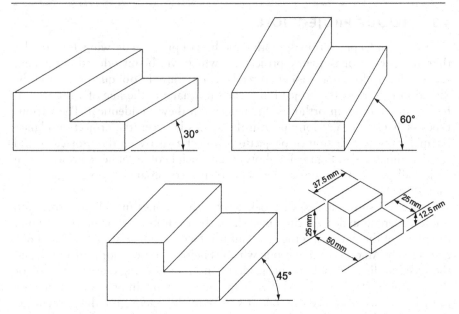

Figure 4.9 Oblique projection without reduction in lengths at the oblique angle.

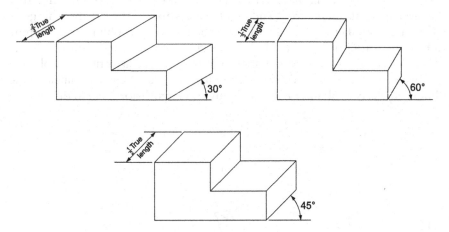

Figure 4.10 Oblique projection with reduction in lengths at the oblique angle.

length is thus altered to $\frac{1}{3}$ × the true length. An oblique angle of 30° causes less distortion and the oblique length is only altered to $\frac{2}{3}$ × the true length. At 45° the true length is reduced by half. These alterations need not be rigidly adhered to. The ones illustrated are chosen because they produce a reasonably true-to-life picture of the block, but a complicated component might have to be drawn with no reduction at all in order to show all the details clearly.

If an oblique drawing is made without any reduction in oblique length, this is sometimes known as 'cavalier projection'. If a reduction in oblique length is made, this is referred to as 'cabinet projection'.

If you were now asked to draw an object in oblique projection, you would probably be very confused when trying to decide which angle to choose and what reduction to make on the oblique lines. If you are asked to produce an oblique drawing, *draw at an oblique angle of 45° and reduce all your oblique dimensions by half, unless you are given other specific instructions.*

Circles and Curves in Oblique Projection

Oblique projection has one very big advantage over isometric projection. Since the front face is drawn in the plane of the paper, any circles on this face are true circles and not ellipses as was the case with isometric projection. Figure 4.11 shows an oblique drawing of a bolt. If the bolt had been drawn in isometric projection, it would have been a long and tedious drawing to make.

There are occasions when there are curves or circles on the oblique faces. When this arises, they may be drawn using the ordinate method that was used for circles on isometric drawings. If the oblique length has been scaled down, then the ordinates on the oblique lengths must be scaled down in the same proportions. Figure 4.12 shows an example of this.

In this case, the oblique angle is 45° and the oblique scale is ½ normal size. The normal 6 mm ordinates are reduced to 3 mm on the oblique faces and the 3 mm ordinates are reduced to 1.5 mm.

It is also worth noting that the ordinates are spaced along a 45° line. This must always be done in oblique projection in order to scale the distances between the ordinates on the oblique view to half those on the plane view.

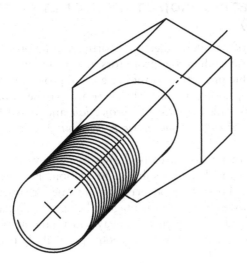

Figure 4.11 Oblique projection with front-facing circles.

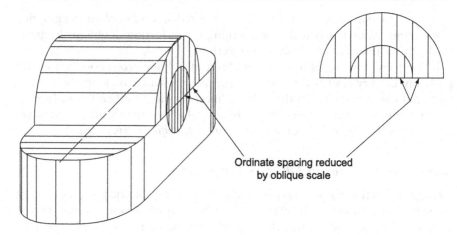

Ordinate spacing reduced
by oblique scale

Figure 4.12 Oblique projection with circles on oblique faces.

The advantage of oblique projection over other pictorial projections is that circles drawn on the front face are not distorted. Unfortunately, examiners usually insist that circles are drawn on the oblique faces, as in Figure 4.12. However, if you are free of the influence of an examiner and wish to draw a component in oblique projection, it is obviously good sense to ensure that the face with the most circles or curves is the front face.

Figure 4.13 shows a small stepped pulley drawn twice in oblique projection. It is obvious that the drawing on the left is easier to draw than the one on the right.

4.4 AXONOMETRIC PROJECTION (ISOMETRIC AND DIMETRIC)

Another method of parallel projection is orthographic projection. In contrast to oblique projection where the projectors are at any angle other than 0° or 90°, in orthographic projection (ortho = right) all projectors are perpendicular to the plane of projection (Figure 4.14). There are two methods of orthographic projection. One is called *multi-view projection* and is used for engineering drawings. The other one, *axonometric projection*, is used to create pictorial drawings.

Engineering drawings are usually drawn in multi-view projection, whose principles are explained in the subsequent section. For the presentation of detailed drawings, this system has been found to be far superior to all others. The system has, however, the disadvantage of being very difficult to understand by people not trained in its usage. It is always essential that an engineer be able to communicate his or her ideas to anybody, particularly people who are not

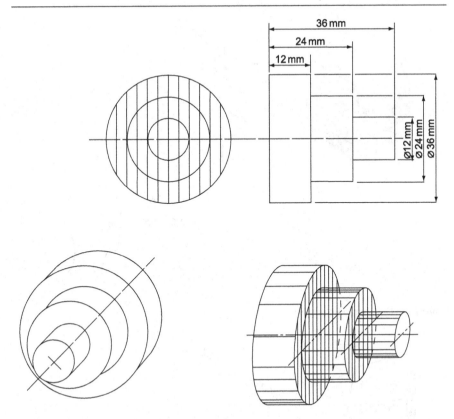

Figure 4.13 A drawing of a stepped pulley.

engineers, and it is therefore an advantage to be able to draw using a system of projection that is more easily understood such as an axonometric projection.

An axonometric projection is a projection method where a pictorial view is created by rotating the object on one of its axes, relative to the projection plane. In this way all three dimensions can be shown. However, not all of them are shown in true size (Figure 4.15).

Axonometric projections can be further classified by the angles between the axonometric axes. When all angles are equal, the projection is classified as *iso-metric* projection. When two of the three angles are equal, the projection is referred to as *dimetric*. When none of the three angles differ in size, the projection is called *trimetric*. The different axonometric projection methods will be further explained in detail.

The so-called 'true isometric' projection is an application of orthographic projection and is dealt with in greater detail later in this chapter. The most common form of isometric projection is called 'conventional isometric'. The isometric axonometric projection method is standardised in ISO 5456–3.

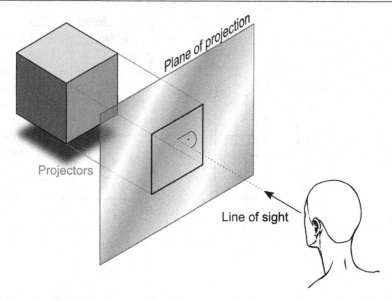

Figure 4.14 Orthographic projection.

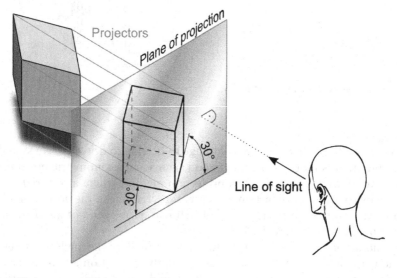

Figure 4.15 Axonometric projection (isometric).

4.4.1 *Conventional Isometric Projection (Isometric Drawing)*

If you were to make a freehand drawing of a row of houses, the house farthest away from you would be the smallest house on your drawing. This is called the perspective of the drawing and, in a perspective drawing, none of the lines

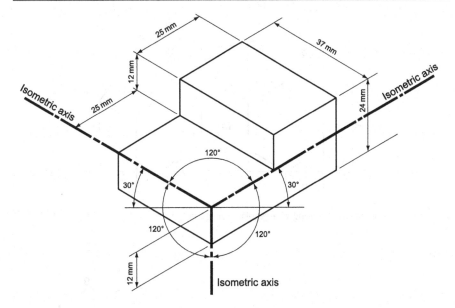

Figure 4.16 Shaped block drawn in conventional isometric projection.

are parallel. Isometric drawing ignores perspective altogether. Lines are drawn parallel to each other and drawings can be made using a T-square and a set square. This is much simpler than perspective drawing.

Figure 4.16 shows a shaped block drawn in conventional isometric projection.

You will note that there are three isometric axes. These are inclined at 120° to each other. One axis is vertical and the other two axes are therefore at 30° to the horizontal. Dimensions measured along these axes, or parallel to them, are true lengths.

The faces of the shaped block shown in Figure 4.16 are all at 90° to each other. The result of this is that all of the lines in the isometric drawing are parallel to the isometric axes. If the lines are not parallel to any of the isometric axes, they are no longer true lengths. An example of this is shown in Figure 4.17 which depicts an isometric drawing of a regular hexagonal prism. The hexagon is first drawn as a plane figure and a simple shape, in this case a rectangle, is drawn around the hexagon. The rectangle is easily drawn in isometric projection and the positions of the corners of the hexagon can be transferred from the plane figure to the isometric drawing with a pair of dividers.

The dimensions of the hexagon should all be 25 mm, and you can see from Figure 4.17 that lines not parallel to the isometric axes do not have true lengths.

Figure 4.18 shows another hexagonal prism. This prism has been cut at an incline, and this means that two extra views must be drawn so that sufficient information to draw the prism in isometric projection can be transferred from the plane views to the isometric drawing.

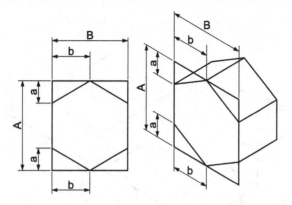

Figure 4.17 An isometric drawing of a regular hexagonal prism.

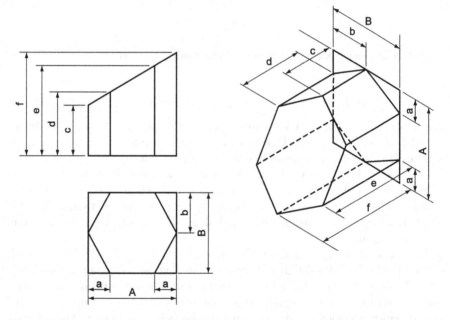

Figure 4.18 A hexagonal prism cut obliquely.

This figure shows that, when making an isometric drawing, all dimensions must be measured parallel to one of the isometric axes.

4.4.2 Circles and Curves Drawn in Isometric Projection

All of the faces of a cube are square. If a cube is drawn in isometric projection, each square side becomes a rhombus. If a circle is drawn on the face of a cube, the circle will change shape when the cube is drawn in isometric projection. Figure 4.19 shows how to plot the new shape of the circle.

The circle is first drawn as a plane figure, and is then divided into an even number of equal strips. The face of the cube is then divided into the same number of equal strips. Centre lines are added and the measurement from the centre line of the circle to the point where strip 1 crosses the circle is transferred from the plane drawing to the isometric drawing with a pair of dividers. This measurement is applied above and below the centre line. This process is repeated for strips 2, 3, etc.

The points that have been plotted should then be carefully joined together with a neat freehand curve.

Figure 4.20 illustrates how this system is used in practice.

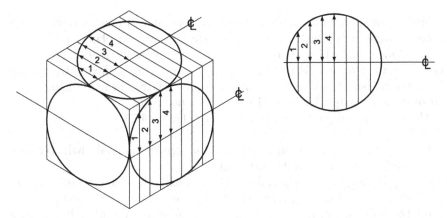

Figure 4.19 Circles in isometric projection.

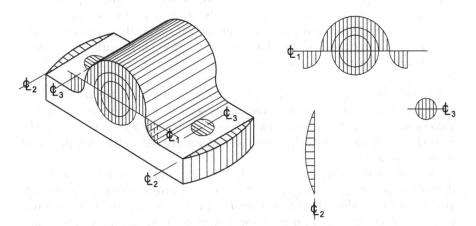

Figure 4.20 How circles and part circles can be drawn isometrically.

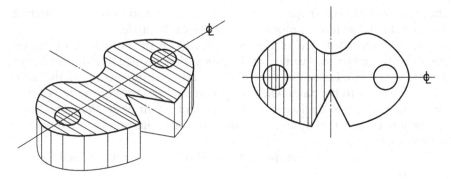

Figure 4.21 A shaped plate drawn in isometric projection.

Since a circle can be divided into four symmetrical quadrants, it is really necessary to draw only a quarter of a circle instead of a whole plane circle.

The dimensions which are transferred from the plane circle to the isometric view are called 'ordinates', and the system of transferring ordinates from plane figures to isometric views is not confined to circles. It may be used for any regular or irregular shape. Figure 4.21 shows a shaped plate.

There are several points worth noting from Figure 4.21.

(a) Since the plate is symmetrical about its centre line, only half has been divided into strips on the plane figure.

(b) In proportion to the plate, the holes are small. They have, therefore, ordinates much closer together so that they can be drawn accurately.

(c) The point where the vee cut-out meets the elliptical outline has its own ordinate so that this point can be transferred exactly to the isometric view.

(d) Since the plate has a constant thickness, the top and bottom profiles are the same. A quick way of plotting the bottom profile is to draw several vertical lines down from the top profile and, with dividers set at the required thickness of the plate, follow the top curve with the dividers, marking the thickness of the plate on each vertical line.

It is sometimes necessary to draw circles or curves on faces that are not parallel to any of the three isometric axes. Figure 4.22 shows a cylinder cut at 45°. Two views of the cylinder have to be drawn: a plan view and an elevation. The plan view is divided into strips, and the positions of these strips are projected onto the elevation.

The base of the cylinder is drawn in isometric projection in the usual way. Points 1 to 20, where the strips cross the circle, are projected vertically upwards and the height of the cylinder, measured from the base with dividers, is transferred for each point in turn from the elevation to the isometric view. These points are then carefully joined together with a neat freehand curve.

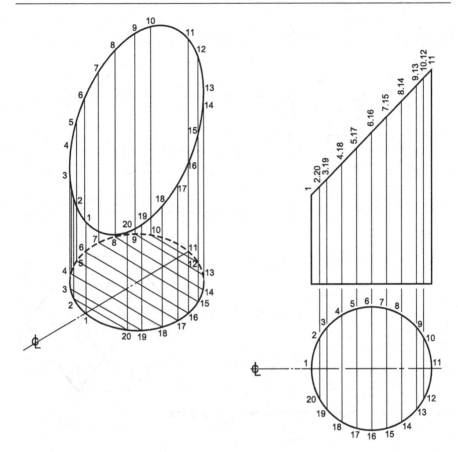

Figure 4.22 A cylinder cut obliquely.

4.4.3 True Isometric Projection

Isometric projection is a method of drawing with instruments which gives a pictorial view of an object. It is not often used in industry and, when it is, the vast majority of drawings are made using conventional isometric projection. Conventional isometric projection is a distorted and simplified form of true isometric. True isometric is found by taking a particular view from an orthographic projection of an object. Figure 4.23 shows a cube, about 25 mm side, drawn in orthographic projection with the cube so positioned that the front elevation is a true isometric projection of the cube. The three isometric axes are still at 120° to each other. In conventional isometric projection, distances measured parallel to these axes are true lengths. In true isometric projection, they are no longer true lengths although they are proportional to their true lengths. However, the horizontal distances on a true isometric projection *are* true lengths. The reduction of lengths measured parallel to the isometric axes makes the overall size of the true isometric drawing appear to look more

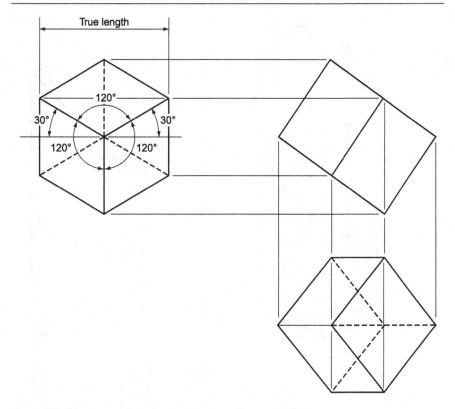

Figure 4.23 True isometric projection (drawn orthographically).

natural, particularly when directly compared with an orthographic or plane view of the same object (compare the relative sizes of the prism in Figure 4.17).

If the horizontal length and the length parallel to the isometric axes were both to be true lengths, the isometric axes would have to be at 45° (Figure 4.24). Since the isometric axes are at 30°, the 45° lengths must be reduced.

This operation is shown in Figure 4.25.

The ratio between the true length and the isometric length is isometric length = true length × 0.8165.

This ratio is constant for all lines measured parallel to any of the isometric axes. If you are asked to draw an object using an isometric scale, your scale may be constructed as in Figure 4.25 or you may construct a conventional plain scale as shown in Figure 4.26. The initial length of this scale is 100 × 0.8165 = 81.65 mm. The scale is then completed as shown in Chapter 2.

Taking the cube from the example shown in Figure 4.19, the isometric representation gives the same visual importance to all three available faces. When one of these faces is of main importance, it is more convenient to use the dimetric axonometry.

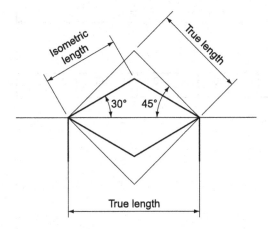

Figure 4.24 Isometric length and true length.

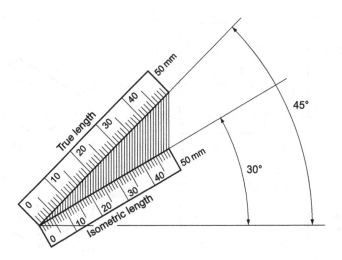

Figure 4.25 Difference in isometric length and true length.

4.4.4 Dimetric Projection

It was shown that in isometric projection the angles between the projection plane and the three coordinate axes X, Y and Z are equal (Greek: *isos* 'equal, identical' + *metron* 'a measure'). Further, the ratio of the three scales is 1:1:1, meaning that all dimensions shown in the direction of any of the three axes are given in true size. In dimetric projection (Greek: *di* 'two' + *metron* 'a measure') just two of the three angles between projection plane and coordinate axes are equal. As a result, the ratio of the three scales is ½:1:1, i.e. a reduction of size in the direction of one coordinate axis, depending on the orientation of

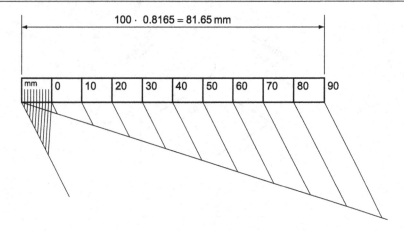

Figure 4.26 A plain scale to measure true isometric lengths.

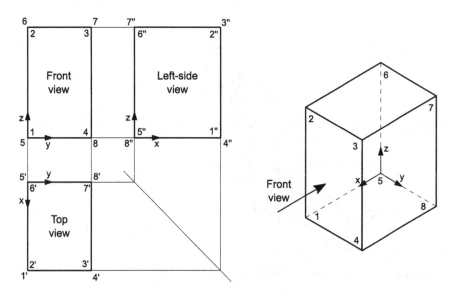

Figure 4.27 Rectangular prism given in three orthogonal views and as a dimetric projection.

the coordinate system. If the three angles and ratios are all different, then the axonometric projection is referred to as 'trimetric' projection.

Figures 4.27–4.29 depict how a dimetric projection is derived from a rectangular prism given in three orthogonal views. For clarity, the corners are numbered and a coordinate system shows the actual orientation of the prism.

As a first step, the prism is rotated in top view by a given angle, as an example by 20° (Figure 4.28). Subsequently, the resulting front view is tilted by the same angle again, that is 20° (Figure 4.29). The left-side view given in Figure 4.29 shows

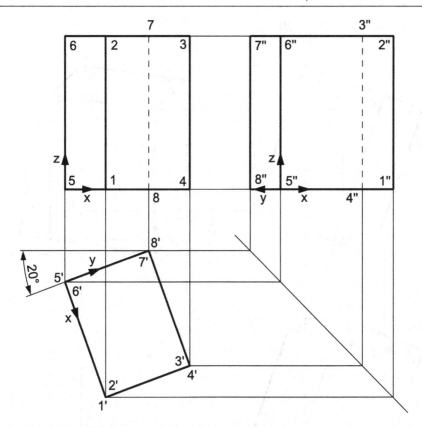

Figure 4.28 Rectangular prism rotated in front view to derive a dimetric projection.

the dimetric representation of the prism. Note that in this view all edges describing the height of the prism appear vertical (original z-direction). In fact they do not show the true length. The edges describing the width of the prism (original y-direction) are slanted in the projection plane by 42° whereas the x-axis, which indicates the direction describing the depth of the prism, is slanted by 7°.

The resulting foreshortening of the edges which we can observe in the derived dimetric representation is usually not considered. When drawing directly a dimetric projection, i.e. without the presented steps, the size reductions are neglected. While the edges in the y- and z-directions would be scaled by a factor of 0.94, they are drawn at full size. The edges in the x-direction would be theoretically reduced in length by a factor of 0.47 but are drawn for convenience at a scale of 0.5.

4.5 MULTI-VIEW PROJECTION (FIRST ANGLE AND THIRD ANGLE)

Multi-view projection is the solution to the biggest problem that a draftsperson has to solve – how to draw, with sufficient clarity, a three-dimensional object

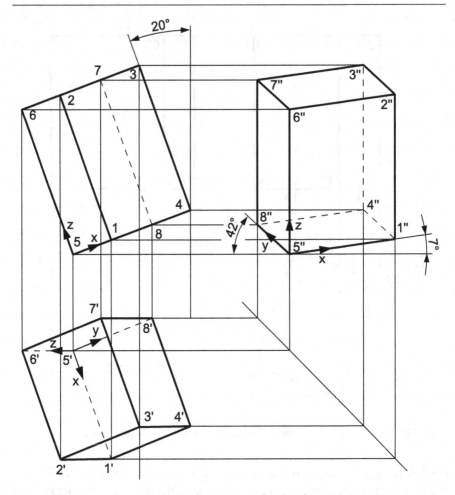

Figure 4.29 Rectangular prism tilted in front view to derive a dimetric projection.

on a two-dimensional plane. The drawing must show quite clearly the detailed outlines of all the faces, and these outlines must be fully dimensioned. If the object is very simple, this may be achieved with a freehand sketch. A less simple object could be drawn in either isometric or oblique projections, although both these systems have their disadvantages. Circles and curves are difficult to draw in either system and neither shows more than three sides of an object in any one view. Multi-view projection, because of its flexibility in allowing any number of views of the same object, has none of these drawbacks.

According to ISO 5456–2, multi-view projection is a special type of ortho-graphic projection and has two forms: first angle and third angle; we shall discuss both.

Figure 4.30 shows a stepped block suspended between two planes.

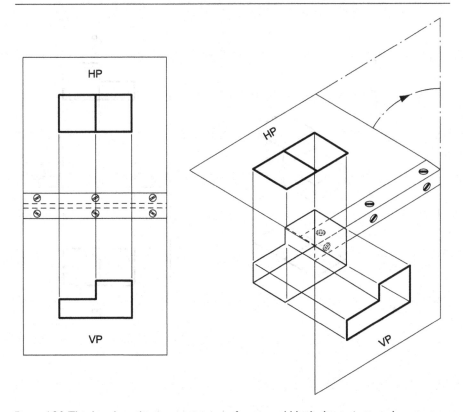

Figure 4.30 Third angle multi-view projection of a stepped block shown in two planes.

A plane is a perfectly flat surface. In this case one of the planes is horizontal and the other is vertical. The view looking on the top of the block is drawn directly above the block on the horizontal plane (HP). The view looking on the side of the block is drawn directly in line with the block on the vertical plane (VP). If you now take away the stepped block and, imagining that the two planes are hinged, fold back the HP so that it lines up with the VP, you are left with two drawings of the block. One is a view looking on the top of the block, and this is directly above another view looking on the side of the block. These two views are called 'elevations'.

Figure 4.30 shows the block in third angle multi-view projection. The same block is drawn in Figure 4.31 in first angle multi-view projection. You still have a VP and a HP but they are arranged differently. The block is suspended between the two planes, and the view of the top of the block is drawn on the HP and the view of the side is drawn on the VP. Again, imagining that the planes are hinged, the HP is folded down so that the planes are in line. This results in the drawing of the side of the block being directly above the drawing of the top of the block (compare this with the third angle drawings).

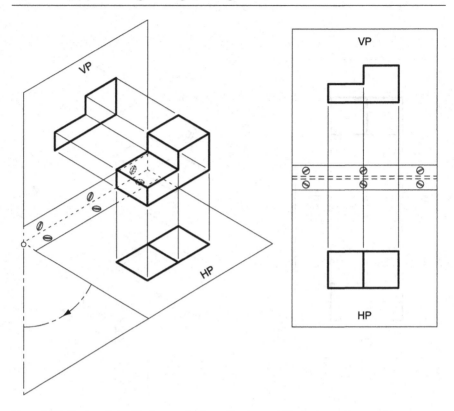

Figure 4.31 First angle multi-view projection of a stepped block shown in two planes.

The reason these two systems are called first and third angle is shown in Figure 4.32. If the HP and the VP intersect as shown, four quadrants are produced. The first quadrant, or first angle, is the top right and the third is the bottom left. If the block is suspended between the VP and the HP in the first and third angles, you can see how the views are projected onto the two planes.

So far we have obtained only two views of the block, one on the VP and one on the HP. With a complicated block this may not be enough. This problem is easily solved by introducing another plane. In this case it is a VP and it will show a view of the end of the block and so, to distinguish it from the other VP, it is called the end vertical plane (EVP), and the original vertical plane is called the front vertical plane (FVP).

The EVP is hinged to the FVP, and when the views have been projected onto their planes, the three planes are unfolded; three views of the block are shown in Figure 4.33.

The drawing on the FVP is called the front elevation (FE) or front view, the drawing on the EVP is called the end elevation (EE) or side view, and the drawing on the HP is called the plan or simply top view. All three views are linked together: the plan is directly above the FE; the EE is horizontally in line with

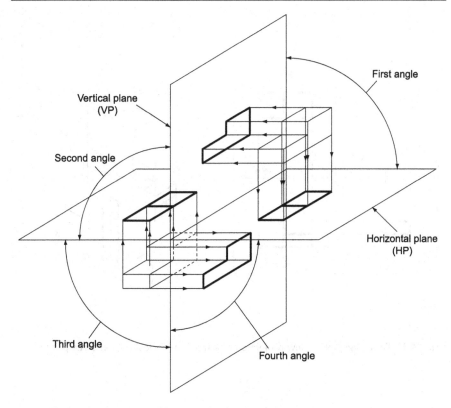

Figure 4.32 Relative positions of first and third angle projections.

the FE; and the plan and the EE can be linked by drawing 45° projection lines. This is why multi-view orthographic projection is so important; it is not just because several views of the same object can be drawn, *it is because the views are linked together*.

Figure 4.33 shows three views of the block drawn in third angle; Figure 4.34 shows three views of the same block drawn in first angle.

In this case the FE is above the plan and to the left of the EE (compare this with third angle). Once again, the EE and the plan can be linked by projection lines drawn at 45°.

The system of suspending the block between three planes and projecting views of the block onto these planes is the basic principle of orthographic projection and must be completely understood if one wishes to study this type of projection any further. This is done in Chapter 5.

The following system is somewhat easier to understand and will meet most of the reader's needs.

Figure 4.35 shows the same shaped block drawn in third angle projection. First, draw the view obtained by looking along the arrow marked FE. This gives you the FE. Now look along the arrow marked EE1 (which points from

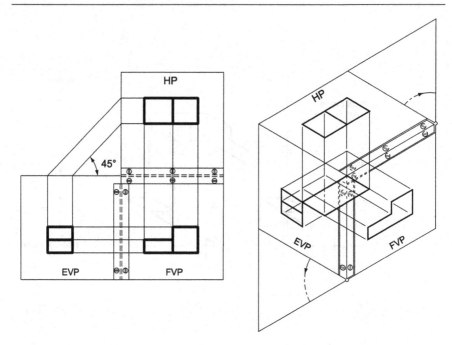

Figure 4.33 Third angle multi-view projection of a stepped block shown in three planes.

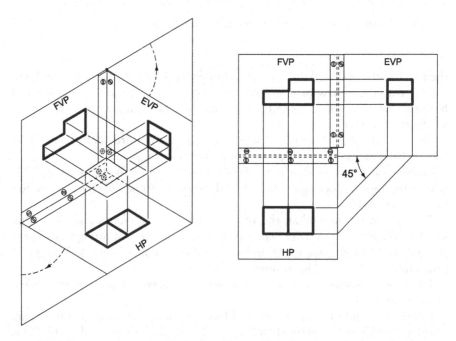

Figure 4.34 First angle multi-view projection of a stepped block shown in three planes.

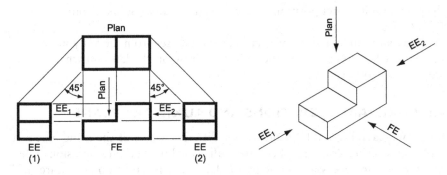

Figure 4.35 Alternative method to develop a third angle multi-view projection.

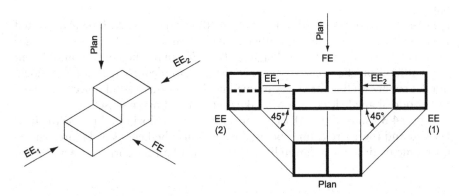

Figure 4.36 Alternative method to develop a first angle multi-view projection.

the left) and draw what you see to the left of the FE. This gives you an EE. Now look along the arrow marked EE2 (which points from the right) and draw what you see to the right of the FE. This gives you another EE. Now look down onto the block, along the arrow marked 'plan' and draw what you see above the FE. This gives the plan, and its exact position is determined by drawing lines from one of the EEs at 45°.

Note that with third angle projection, what you see from the left you draw on the left, what you see from the right you draw on the right and what you see from above you draw above.

Figure 4.36 shows the same block drawn in first angle projection. Again, first draw the view obtained by looking along the arrow marked FE. This gives the FE. Now look along the arrow marked EE1 (which points from the left) and draw what you see to the right of the FE. This gives you an EE. Now look along the arrow marked EE2 (which points from the right) and draw what you see to the left of the FE. This gives you another EE. Now look down on the block, along the arrow marked 'plan', and draw what you see below the FE. This gives the plan, and its exact position is determined by drawing lines from one of the EEs at 45°.

Note that with first angle projection, what you see from the left you draw on the right, what you see from the right you draw on the left and what you see from above you draw below.

4.6 AUXILIARY ELEVATIONS AND AUXILIARY PLANS

So far we have been able to draw four different views of the same block. In most engineering drawings these are sufficient, but there are occasions when other views are necessary, perhaps to clarify a particular point. Figure 4.37 shows two examples where a view other than a FE or an EE is needed to show very important features of a flanged pipe and a bracket.

These extra elevations are called auxiliary elevations (AE) or auxiliary plans (AP).

Figure 4.38 shows an AE and an AP of the shaped block. One is projected from the plan at 30° and the other from the FE at 45°. Projection lines are drawn at those angles and the heights, H and h, are marked off on one AE and the width W on the other. Remember that we are dealing with a solid block, not flat shapes on flat paper. Try to imagine the block as a solid object and these rather odd-shaped elevations will take on form and make sense.

Figure 4.39 shows two AP of a more complicated block. In this case the base is tilted and therefore cannot be used to measure the heights as before. This is overcome by drawing a datum line. The heights of all the corners are measured

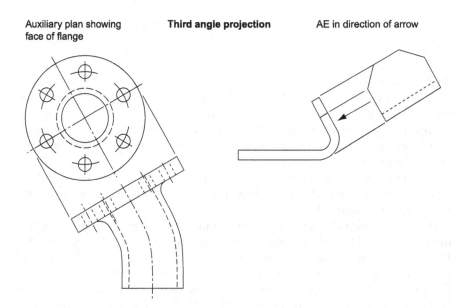

Auxiliary plan showing face of flange   **Third angle projection**   AE in direction of arrow

Figure 4.37 Views of a flanged pipe and a bracket.

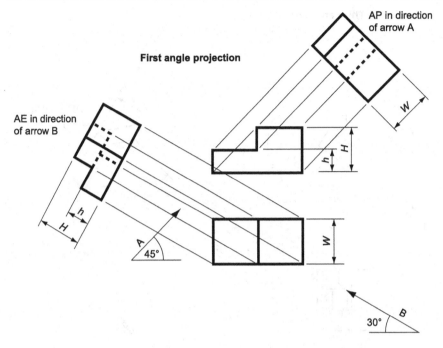

Figure 4.38 Views of a shaped block.

from this datum. Note that on the AP the datum is drawn at 90° to the projection lines.

If the outline contains circles or curves, the treatment is similar. Select some points on the curve and mark off their distances from some convenient datum. In Figure 4.40 this gives dimensions *a*, *b*, *c*, *d*, *e* and *f*. The positions of these points are marked on the plan and they are projected onto the AE. The dimensions *a* to *f* are then marked off on the AE and the points joined together with a neat freehand curve.

It is worth stating again the difference between first and third angle projection, particularly if checked against the previous examples. With first angle, if you look from one side of a view you draw what you see on the *other* side of that view. With third angle, if you look from one side of a view you draw what you see on the *same* side of that view.

Shown next are some of the more common solid geometric solids drawn in orthographic projection.

Prisms and Pyramids

Figure 4.41 shows the following views of a rectangular prism, drawn in first angle projection with the prism tilted at 45° in the FE.

An FE looking along arrow *A*.

An EE looking along arrow *B*.

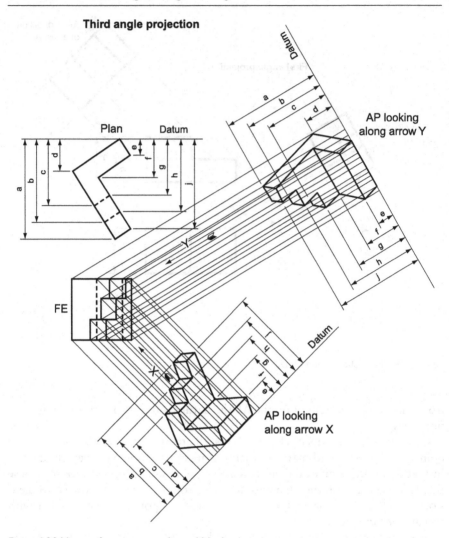

Figure 4.39 Views of a more complicated block.

A plan.

An AP showing the cross-sectional shape of the prism.

Figure 4.42 shows the following views of a square prism drawn in third angle projection. The top of the prism has been cut obliquely at 30°.

An FE looking along arrow *A*.

An EE looking along arrow *B*.

A plan.

An AP projected from the FE at 30°.

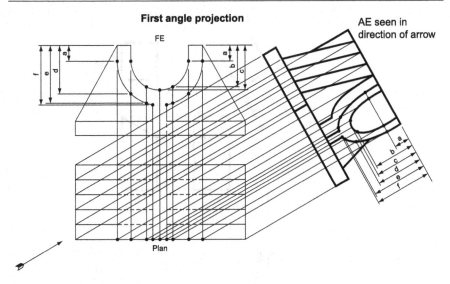

Figure 4.40 Development of an auxiliary view containing curves.

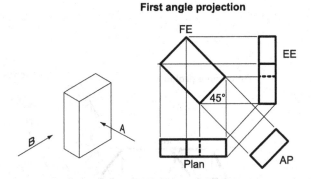

Figure 4.41 Views of a rectangular prism.

Figure 4.43 shows the following views of a regular hexagonal prism, drawn in third angle projection with the prism tilted at 30° in the FE. The top of the prism has been cut obliquely at 45°.

An FE looking along arrow *A*.

An EE looking along arrow *B*.

A plan.

The first view that is drawn is the AP. This is not in the instructions but without it the FE is very difficult to draw. Arrow A indicates that three sides of the hexagon are seen in the FE and the AP is constructed so that three sides are seen (rotate the hexagon through 30° in the AE and only two sides are seen). The AP is also used to find the width of the prism in the EE.

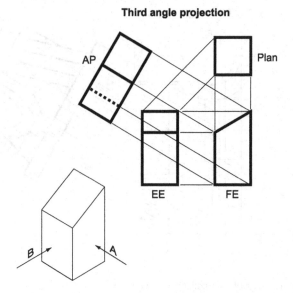

Figure 4.42 Views of a square prism.

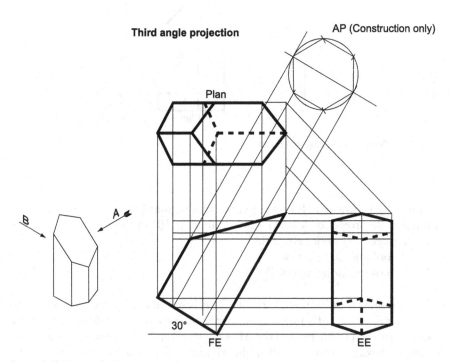

Figure 4.43 Views of a hexagonal prism.

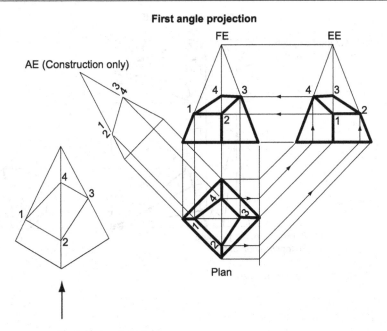

Figure 4.44 Views of a square pyramid.

Figure 4.44 shows the following views of the frustum of a square pyramid drawn in first angle projection. The corners of the pyramid are numbered 1 to 4 for easy identification on each elevation.

An FE looking along the arrow.

An EE seen from the left of the FE.

A plan.

With this type of problem it is wise initially to draw the required views as if the pyramid were complete. Once again it is necessary to draw an AE so that the oblique face can be drawn on the AE and then points 1, 2, 3 and 4 can be projected back onto the plan. Points 1 and 3 are then projected onto the FE and points 2 and 4 onto the EE. Points 2 and 4 can be projected from the EE to the FE and points 1 and 3 from the FE to the EE. Note that once the AE has been drawn it is possible to draw the oblique face on all three views without any further measuring.

Figure 4.45 shows the following views of an octagonal pyramid drawn in third angle projection. The pyramid is lying on its side.

An FE looking along the arrow.

An EE seen from the right of the FE.

A plan.

To draw the pyramid lying on its side, first draw it standing upright and then tip it over. This is done with compasses as shown. If a plan of the pyramid standing upright is constructed, it makes it easier to find the positions of the corners of the pyramid in the plan when it has been tipped over.

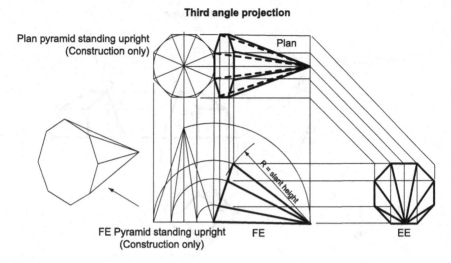

Figure 4.45 Views of an octagonal prism.

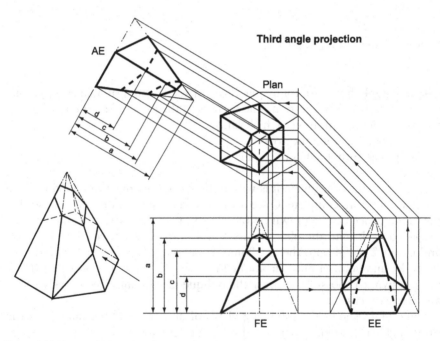

Figure 4.46 Views of a hexagonal pyramid.

Figure 4.46 shows the following views of a hexagonal pyramid drawn in third angle projection. The top of the pyramid is cut at 45° and the bottom at 30°.
An FE seen in the direction of the arrow.
An EE seen from the right of the FE.

A plan.

An AE projected from the plan at 30°.

As for Figure 4.44, the pyramid is first drawn as if it were complete, on all four views. The lower cutting plane is then drawn on the FE. The points where it crosses the corners are then projected across to the EE and up to the plan. The point where it crosses the centre corner on the FE cannot be projected straight to the plan and has to be projected via the EE (follow the arrows).

The upper cutting plane is then drawn on the EE, and the points where it crosses the corners are projected across to the FE and up to the plan.

Most of these corners can be projected straight from the plan onto the AE. The exceptions are the points on the centre corner and these (dimensions a, b, c and d) can be transferred from any convenient source, in this case the FE.

Cylinders and Cones

Figure 4.47 shows the following views of a cylinder drawn in first angle projection.

An FE seen in the direction of the arrow.

A plan.

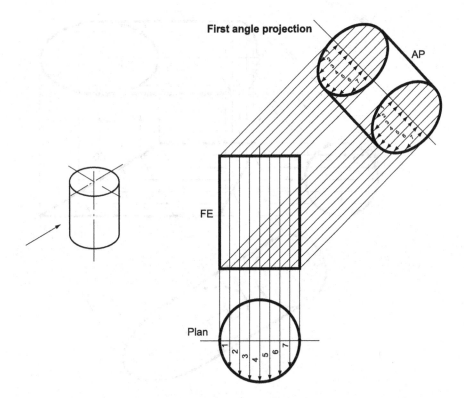

Figure 4.47 Views of a cylinder drawn in first angle projection.

An AP projected from the FE at 45°.

If the plan is divided into several strips, the width of the cylinder at any one of these strips can be measured. The exact positions of each of the strips can be projected onto the FE and then across to the AP. The width of the cylinder at each of the strips is transferred from the plan onto the AP with dividers, measured each side of the centre line (only one side is shown). The points are then joined together with a neat freehand curve.

Figure 4.48 shows the following views of a cylinder drawn in third angle projection. The cylinder is lying on its side and one end is cut off at 30° and the other end at 60°.

An FE seen in the direction of the arrow.

An EE seen from the left of the FE.

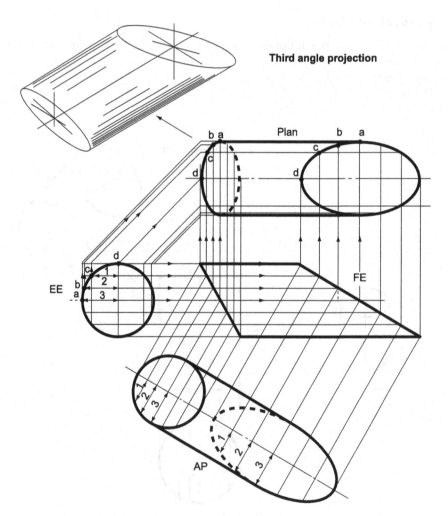

Figure 4.48 Views of a cylinder drawn in third angle projection.

A plan.

An AP projected from the plan at 60°.

The EE is divided into several strips. The strips are projected from the EE to the FE and up to the plan. They are also projected from the EE to the plan at 45°. The points where the projectors from the FE and the EE meet on the plan (at *a*, *b*, *c* and *d*, etc.) give the outline of the two ellipses on the plan.

The outline of the ellipses on the AP are found by projecting the strips onto the AP and then transferring measurements 1, 2, 3, etc., from the EE to the AP with dividers.

Figure 4.49 shows the following views of a curved cylinder drawn in first angle projection.

An FE seen in the direction of the arrow.

An EE seen from the left of the FE.

A plan.

An AP projected from the FE at 45°.

This drawing uses a different method of plotting an AP from the previous two examples. Instead of being divided into strips, the cylinder is divided into 12 equal segments. These are marked on the walls of the cylinder as numbers, from 1 to 12. The ellipses formed on the AP are found by plotting the

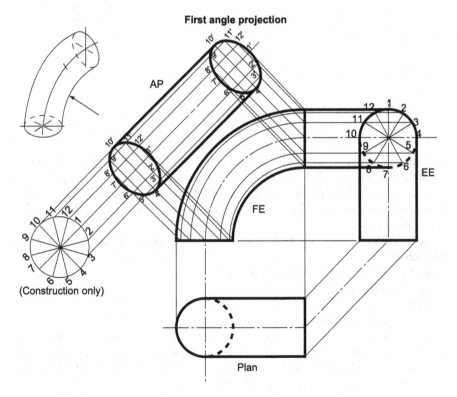

Figure 4.49 Views of a curved cylinder.

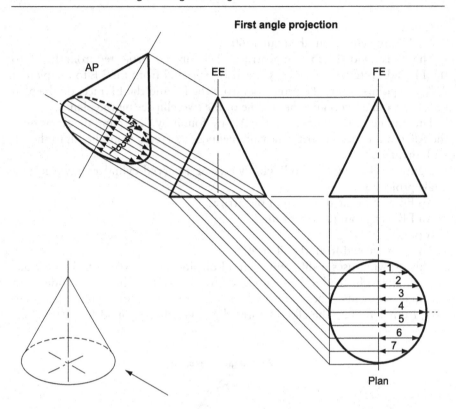

Figure 4.50 Views of a cone.

intersections of the projectors of numbers 1 to 12 from the FE and from a construction drawn in line with the AP. The projectors intersect in 1', 2', 3' etc. Note that on the EE number 1 is at the top of the circle while on the construction (and hence on the AP) number 1 is on the right. This, of course, is what you should expect.

Figure 4.50 shows the following views of a cone drawn in first angle projection.

An FE seen in the direction of the arrow.

An EE seen from the right of the FE.

A plan.

An AP projected from the EE at 30°.

The plan is divided into strips. These strips are projected across to the EE and hence to the AP. The width of the base of the cone on each of these strips is measured on the plan with dividers and transferred onto the AP. The points are then joined with a neat freehand curve.

4.7 PROJECTION OF SECTIONED SOLIDS

4.7.1 Section Planes

Suppose that you make a drawing of a box. You draw the box in orthographic projection and are pleased with the result. But someone comes along and says, quite reasonably, 'It's a good drawing but, after all, a box is only a container and you haven't shown what is inside the box; surely that is what is important'. And of course, he is right.

It is often vital to show what is inside an object as well as to show the outside. In orthographic projection, this is catered for with a section.

Figure 4.51 shows two drawings of the same block, one drawn without a section and one drawn with a section. The upper drawing does not show clearly on any one of the orthographic views that the block is hollow. On the lower left isometric view, the block has been cut in half and it is immediately obvious

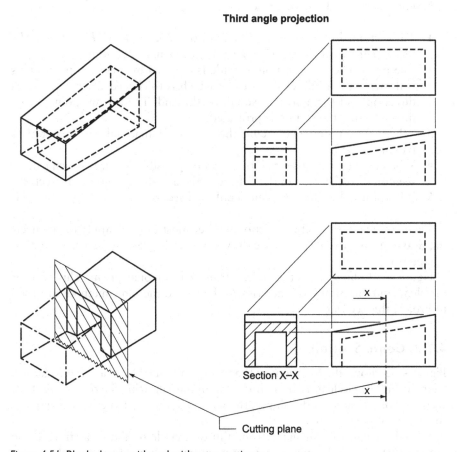

Third angle projection

Section X–X

Cutting plane

Figure 4.51 Block shown with and without a section.

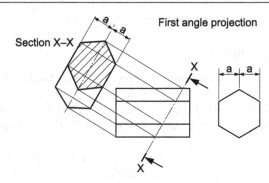

Figure 4.52 Hexagonal prism with a section.

that the block is hollow. The lower right view shows the cut block drawn in orthographic projection. Again, it is much easier to see that the block is hollow.

Note carefully the following rules:

1   The sectioned EE is drawn with half of the block missing, *but none of the other views are affected.* They keep their normal full outline.
2   The point where the section is made is denoted by a cutting plane. This is drawn with a thin chain dot line which is thickened where it changes direction and for a short distance at the end. The arrows point in the direction that the section is projected.
3   Where the cutting plane cuts through solid material, the material is hatched at 45°.
4   When a section is projected, the remaining visible features which can be seen on the other side of the cutting plane are also drawn on the section.
5   It is not usual to draw hidden detail on a section.

There are many rules about sectioning, but most of them apply to engineering drawing rather than geometric drawing. For this reason they are found in Chapter 6.

Figure 4.52 shows a section taken from a hexagonal prism. This type of problem contains all the characteristics of AE, and the same methods are used to project this section.

4.7.2 Conic Sections

Figure 4.53 shows the five sections that can be obtained from a cone. So far the triangle and the circle have been discussed; in the following the remaining three sections are introduced: the ellipse, the parabola and the hyperbola. These are three very important curves.

The ellipse can vary in shape from almost a circle to almost a straight line and is often used in designs because of its pleasing shape. The parabola can be seen in the shape of electric fire reflectors, radar dishes and the main cable of

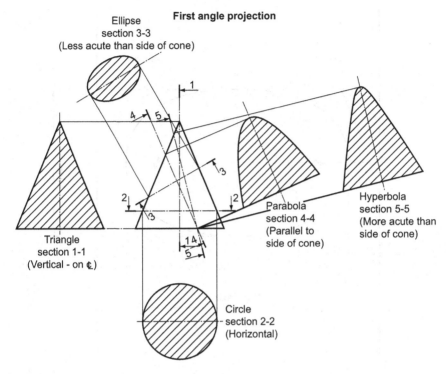

Figure 4.53 Conic sections.

suspension bridges. Both the parabola and the hyperbola are much used in civil engineering. The immense strength of structures that are parabolic or hyperbolic in shape has led to their use in structures made of precast concrete and where large unsupported ceilings are needed.

The height of the cone and the base diameter, with the angle of the section relative to the side of the cone, are the factors that govern the relative shape of any ellipses, parabolas or hyperbolas. There are an infinite number of these curves and, given eternity and the inclination to do so, you could construct them all by taking sections from cones. There are other ways of constructing these curves. This chapter, as well as showing how to obtain them by plotting them from cones, shows some other equally important methods of construction.

The Ellipse

Figure 4.54 shows an ellipse with the important features labelled.

The Ellipse as a Conic Section

Figure 4.55 shows in detail how to project an ellipse as a section of a cone. The shape across X–X is an ellipse.

First, draw the FE, the EE and the plan of the complete cone. Divide the plan into 12 equal sectors with a 60° set square. Project these sectors onto the FE and the EE where they appear as lines drawn on the surface of the cone from

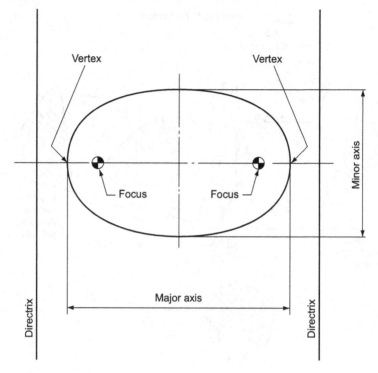

Figure 4.54 The ellipse.

the base to the apex. The points where these lines cross X–X can be easily projected across to the EE and down to the plan to give the shape of X–X on these elevations. The point on the centre line must be projected onto the plan via the EE (follow the arrows).

The shape of X–X on the plan is not the true shape since, in the plan, X–X is sloping down into the page. However, the widths of the points, measured from the centre line, are true lengths and can be transferred from the plan to the auxiliary view with dividers to give the true shape across X–X. This is the ellipse.

The Ellipse as a Locus

DEFINITION

An ellipse is the locus of a point that moves so that its distance from a fixed point (called the focus) bears a constant ratio, always less than 1, to its perpendicular distance from a straight line (called the directrix). An ellipse has two foci and two directrices.

Figure 4.56 shows how to draw an ellipse given the relative positions of the focus and the directrix, and the eccentricity. In this case the focus and the directrix are 20 mm apart and the eccentricity is $\frac{3}{4}$.

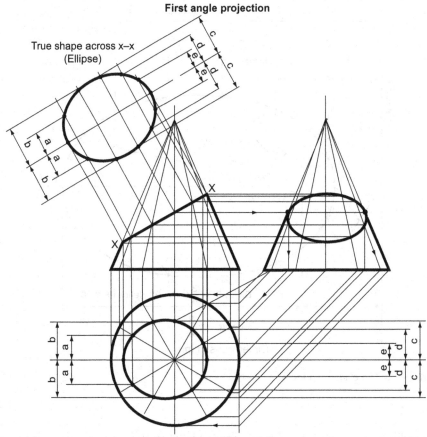

Figure 4.55 The ellipse as a conic section.

The first point to plot is the one that lies between the focus and the directrix. This is done by dividing DF in the same ratio as the eccentricity, 4:3. The other end of the ellipse, point P, is found by working out the simple algebraic sum shown in Figure 4.56.

The condition for the locus is that it is always $\frac{3}{4}$ as far from the focus as it is from the directrix. It is therefore $\frac{4}{3}$ as far from the directrix as it is from the focus. Thus, if the point is 30 mm from F, it is $\frac{40}{3}$ mm from the directrix; if the point is 20 mm from F, it is $\frac{4}{3} \times 20$ mm from the directrix; if the point is 30 mm from F, it is $\frac{3}{4} \times 30$ mm from the directrix. This is continued for as many points as may be necessary to draw an accurate curve. The intersections

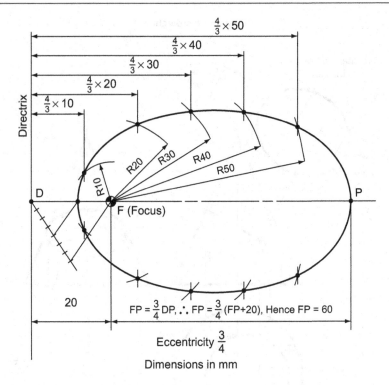

Figure 4.56 The ellipse as a locus.

of radii drawn from F and lines drawn parallel to the directrix, their distance from the directrix being proportional to the radii, give the outline of the ellipse. These points are joined together with a neat freehand curve.

To Construct an Ellipse by Concentric Circles.
We now come to the first of three simple methods of constructing an ellipse. All three methods need only two pieces of information for the construction – the lengths of the major and minor axes.

Stage 1. Draw two concentric circles, radii equal to ½ major and ½ minor axes.

Stage 2. Divide the circle into a number of sectors. If the ellipse is not too large, 12 will suffice as shown in Figure 4.57.

Stage 3. Where the sector lines cross the smaller circle, draw horizontal lines towards the larger circle. Where the sector lines cross the larger circle, draw vertical lines to meet the horizontal lines.

Stage 4. Draw a neat curve through the intersections.

To Construct an Ellipse in a Rectangle.
Stage 1. Draw a rectangle, length and breadth equal to the major and minor axes.

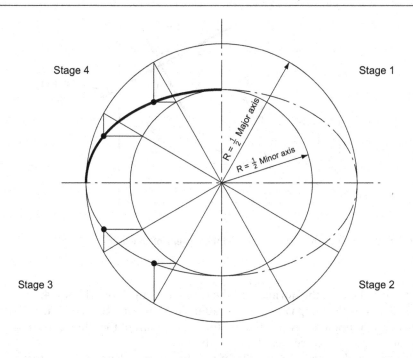

Figure 4.57 To construct an ellipse by concentric circles.

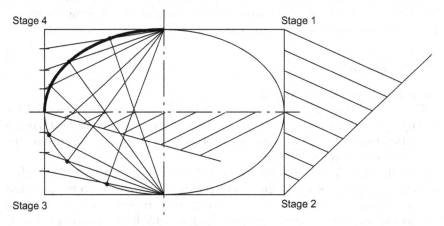

Figure 4.58 To construct an ellipse in a rectangle.

Stage 2. Divide the two shorter sides of the rectangle into the same *even* number of equal parts. Divide the major axis into the same number of equal parts.

Stage 3. From the points where the minor axis crosses the edge of the rectangle, draw intersecting lines as shown in Figure 4.58.

Stage 4. Draw a neat curve through the intersections.

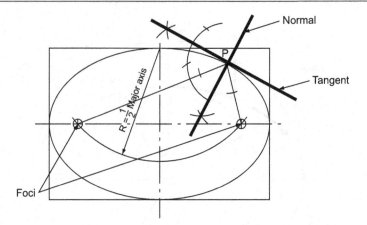

Figure 4.59 To find the foci, the normal and the tangent of an ellipse.

To find the foci, the normal and the tangent of an ellipse (Figure 4.59).

The foci. With compasses set at a radius of ½ major axis, centre at the point where the minor axis crosses the top (or bottom) of the ellipse, strike an arc to cut the major axis twice. These are the foci.

The normal at any point P. Draw two lines from P, one to each focus and bisect the angle thus formed. This bisector is a normal to the ellipse.

The tangent at any point P. Since the tangent and normal are perpendicular to each other by definition, construct the normal and erect a perpendicular to it from P. This perpendicular is the tangent.

The Parabola

The Parabola as a Conic Section

The method used for finding the ellipse in Figure 4.55 can be adapted for finding a parabolic section. However, the method shown in Figure 4.60 is much better because it allows for many more points to be plotted. In Figure 4.60 the shape across Y–Y is a parabola.

First, divide Y–Y into a number of equal parts, in this case six. The radius of the cone at each of the seven spaced points is projected onto the plan and circles are drawn. Each of the points must lie on its respective circle. The exact position of each point is found by projecting it onto the plan until it meets its circle. The points can then be joined together on the plan with a neat curve.

The EE is completed by plotting the intersection of the projectors of each point from the FE and the plan.

Neither the EE nor the plan shows the true shape of Y–Y since, in both views, Y–Y is sloping into the paper. The only way to find the true shape of Y–Y is to project a view at right angles to it. The width of each point, measured from the centre line, can be transferred from the plan as shown.

The Parabola as a Locus

Third angle projection

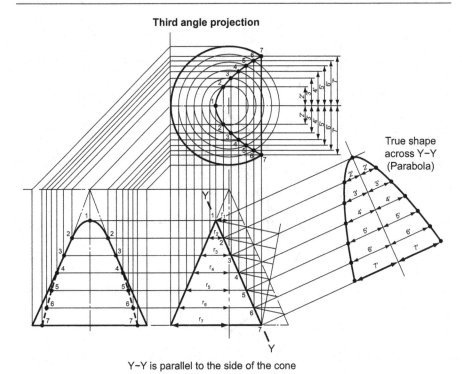

True shape
across Y–Y
(Parabola)

Y–Y is parallel to the side of the cone

Figure 4.60 The parabola as a conic section.

Definition

A parabola is the locus of a point that moves so that its distance from a fixed point (called the focus) bears a constant ratio of 1 to its perpendicular distance from a straight line (called the directrix).

Figure 4.61 shows how to draw a parabola given the relative positions of the focus and the directrix. In this case the focus and directrix are 20 mm apart.

The first point to plot is the one that lies between the focus and the directrix. By definition it is the same distance, 10 mm, from both.

The condition for the locus is that it is always the same distance from the focus as it is from the directrix. The parabola is therefore found by plotting the intersections of radii 15 mm, 20 mm, 30 mm, etc., centre on the focus, with lines drawn parallel to the directrix at distances 15 mm, 20 mm, 30 mm, etc.

The Hyperbola

The Hyperbola as a Conic Section

The method is identical to that used for finding the parabolic section in Figure 4.60. The construction, Figure 4.62, can be followed from the instructions for that figure.

The Hyperbola as a Locus

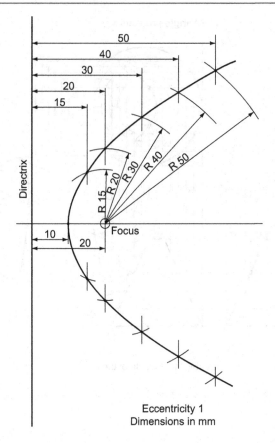

Eccentricity 1
Dimensions in mm

Figure 4.61 The parabola as a locus.

Definition

A hyperbola is the locus of a point that moves so that its distance from a fixed point (called the focus) bears a constant ratio, always greater than 1, to its perpendicular distance from a straight line (called the directrix).

Figure 4.63 shows how to draw a hyperbola given the relative positions of the focus and the directrix (in this case 20 mm) and the eccentricity (3/2).

The first point to plot is the one that lies between the focus and the directrix. This is done by dividing the distance between them in the same ratio as the eccentricity, 3:2.

The condition for the locus is that it is always $\frac{2}{3}$ as far from the directrix as it is from the focus. Thus, if the point is 15 mm from the focus, it is $\frac{2}{3}$ × 15 mm from the directrix; if it is 20 mm from the focus, it is $\frac{2}{3}$ × 20 mm from the directrix. This is continued for as many points as may be required.

Third angle projection

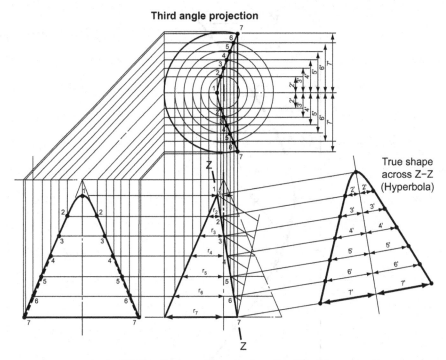

True shape
across Z–Z
(Hyperbola)

Angle of Z–Z to the base is greater than the side of the cone

Figure 4.62 The hyperbola as a conic section.

There are constructions for the normal and tangent to a hyperbola, but they introduce additional features that are beyond the scope of this book.

4.8 PROBLEMS

(All questions originally set in imperial units.)

Oblique Projection

1 Figure 4.64 shows two views of a small casting. Draw, full size, an oblique projection of the casting with face A towards you.
2 Figure 4.65 shows two views of a cast iron hinge block. Make an oblique drawing of this object, with face A towards you, omitting hidden detail.

   North Western Secondary School Examinations Board

3 Figure 4.66 shows the outline of the body of a depth gauge. Make an oblique drawing, twice full size, of the body with corner A towards you.
4 Figure 4.67 shows two views of a holding-down clamp. Draw the clamp, full size, in oblique projection with corner A towards you.

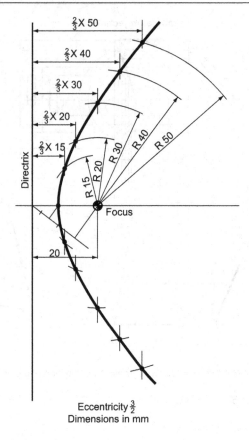

Eccentricity $\frac{3}{2}$
Dimensions in mm

Figure 4.63 The hyperbola as a locus.

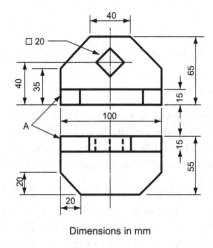

Dimensions in mm

Figure 4.64

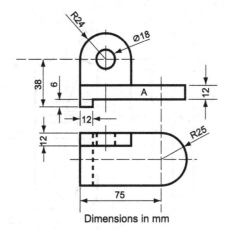

Dimensions in mm

Figure 4.65

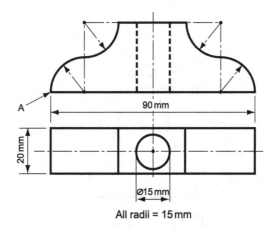

All radii = 15 mm

Figure 4.66

5 Two views of a casting are shown in Figure 4.68. Draw (a) the given views and (b) an oblique pictorial view, looking in the direction shown by arrow *L*, using cabinet projection, that is with the third dimension at 45° and drawn half-size.

Associated Examining Board

6 Two views of a machined block are shown in Figure 4.69. Draw, full size, an oblique projection of the block with AB sloping upwards to the right at an angle of 30°. Use half-size measurements along the oblique lines. The curve CD is parabolic and D is the vertex of the curve. Hidden detail need not be shown.

Oxford Local Examinations

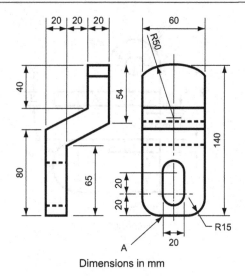

Dimensions in mm

Figure 4.67

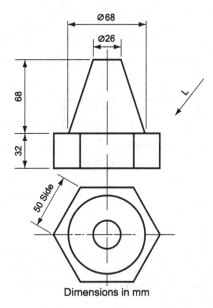

Dimensions in mm

Figure 4.68

7   A special link for a mechanism has dimensions as shown in Figure 4.70.
Draw an oblique view of this link resting on the flat face, using an angle
of 30°, with the centre line marked AB sloping upwards to the right and
with all dimensions, full size. Radius curves may be sketched in and hidden details are to be omitted.

Oxford Local Examinations

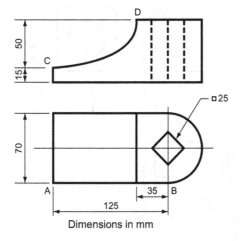

Figure 4.69

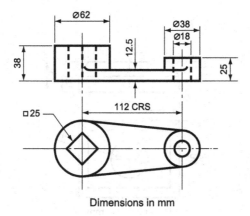

Dimensions in mm

Figure 4.70

Isometric Projection

8 Draw, full size, an isometric projection of the component shown in Figure 4.71 looking in the direction of the arrow A. Hidden details are not to be shown.

Associated Lancashire Schools Examining Board

9 Figure 4.72 shows the front elevation and plan of an ink bottle stand. Make a full size isometric drawing of the stand with corner A nearest to you. Hidden details should *not* be shown.

West Midlands Examinations Board

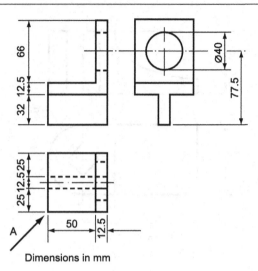

Dimensions in mm

Figure 4.71

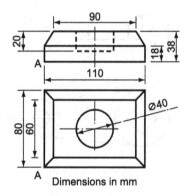

Dimensions in mm

Figure 4.72

10 Figure 4.73 shows the development of a hexagonal box. Draw, in isometric projection, the assembled box standing on its base. Ignore the
thickness of the material and omit hidden detail.

North Western Secondary School Examinations Board (See Chapter 5
for information not in Chapter 4.)

11 Three views of a bearing are shown in Figure 4.74. Make an isometric
drawing of the bearing. Corner A should be the lowest point on your
drawing. No hidden details are required.

South-East Regional Examinations Board

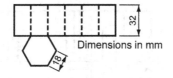

Dimensions in mm

Figure 4.73

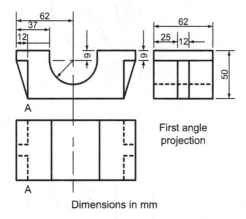

First angle
projection

Dimensions in mm

Figure 4.74

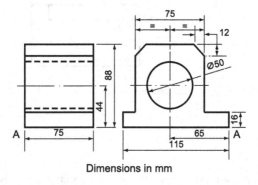

Dimensions in mm

Figure 4.75

12 Two views of a plain shaft bearing are shown in Figure 4.75. Make a full size isometric drawing of the bearing. Hidden details should *not* be shown.

West Midlands Examinations Board

13 Construct an isometric drawing of the casting shown in Figure 4.76. Make point X the lowest point of your drawing. Do not use an isometric scale.

University of London School Examinations

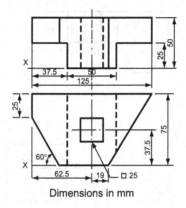

Dimensions in mm

Figure 4.76

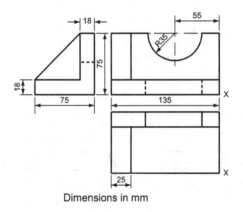

Dimensions in mm

Figure 4.77

14 Construct an isometric scale and use it to make a true isometric view of the casting shown in Figure 4.77. Corner X is to be the lowest corner of your drawing.

University of London School Examinations

15 A plan and elevation of the base of a candlestick are shown in Figure 4.78. Draw (a) another elevation when the base is viewed in the direction of the arrow and (b) an accurate isometric view of that half of the candlestick base indicated by the letters *abcd* on the plan view. The edge *ab* is to be in the foreground of your drawing.

Southern Universities' Joint Board

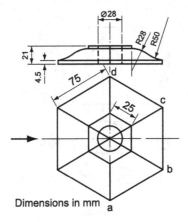

Figure 4.78

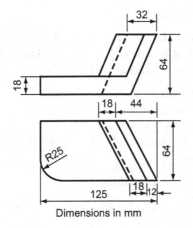

Dimensions in mm

Figure 4.79

16 Figure 4.79 shows the plan and elevation of an angle block. Make a full size isometric projection of the block, making the radiused corner the lowest part of your drawing. Do *not* use an isometric scale. Hidden detail should be shown.

Oxford and Cambridge Schools Examinations Board

17 Figure 4.80 shows two views of a cylindrical rod with a circular hole. Make an isometric drawing of the rod in the direction of the arrows R and S. Hidden detail is not required.

Associated Examining Board

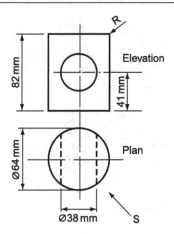

Figure 4.80

Orthographic Projection

18 Figure 4.81 shows the elevation of a 20 mm square prism 50 mm long resting with one of its corners on the HP. Draw, full size, the following views and show all the hidden detail.

(a) The given elevation.
(b) An EE looking in the direction of arrow *E*.
(c) A plan projected beneath view (a).
North Western Secondary School Examinations Board

19 Figure 4.82 shows the FE and an isometric sketch of a sheet metal footlight reflector for a puppet theatre.

Draw, full size, the given FE and from it project the plan and end view of the reflector.
Draw, also, the true shape of the surface marked ABCD.
The thickness of the metal can be ignored.
West Midlands Examinations Board

20 The plan and elevation of a hexagonal distance piece are shown in Figure 4.83. Draw these views, full size, and project an AE on X_1Y_1. Hidden details are not to be shown.

Associated Lancashire Schools Examining Board

21 Figure 4.84 shows details of a cast concrete block. To a scale of 10 mm = 100 mm draw the following:

(a) The two given views.
(b) An EE looking in the direction of the arrow *K*.
(c) The true shape of the sloping surface AB.
Metropolitan Regional Examinations Board

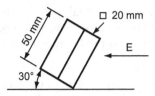

Figure 4.81

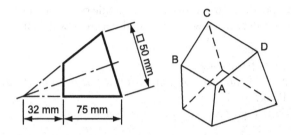

Figure 4.82

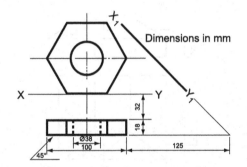

Figure 4.83

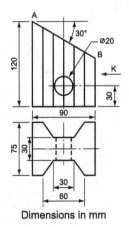

Dimensions in mm

Figure 4.84

22 Two views are shown of a bungalow which has been made as a model (Figure 4.85). To obtain a better impression of its design, a view in the direction of arrow M is required. Draw, full size, the following:

(a) The two given views.
(b) An AE, projected from the plan, in the direction of arrow M.
Metropolitan Regional Examinations Board

23 The front view of Figure 4.86 (drawing A) is of a piece of metal with the left-hand portion bent upwards at an angle of 45° as shown. The bottom drawing is the plan view of the metal before it was bent.

Draw, full size:
(a) the given front view;
(b) the plan of the piece of metal *after it has been bent up*. The curve X can be ignored.
South-East Regional Examinations Board

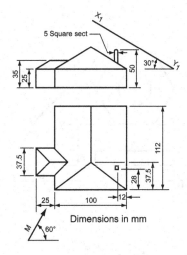

Figure 4.85

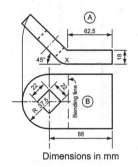

Figure 4.86

24 An elevation of a cone lying on its side is given in Figure 4.87. Copy the given elevation full size and from it project the plan and end view of the cone.

West Midlands Examinations Board

25 An elevation of a machined part is given in Figure 4.88. Draw the following views, full size:

(a) the elevation as shown;
(b) an end view as seen in the direction of arrow *A*;
(c) the true shape of the sloping surface.
South-East Regional Examinations Board

26 Draw the two elevations of the machined section shown in Figure 4.89 and add a plan in the direction of the arrow *P* showing hidden lines. Scale: full size.

Oxford and Cambridge Schools Examination Board

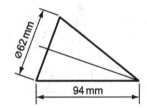

Figure 4.87

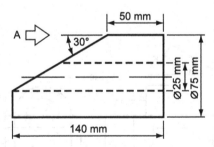

Figure 4.88

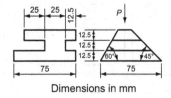

Dimensions in mm

Figure 4.89

Auxiliary Elevations

27 Figure 4.90 shows the elevation of a right hexagonal pyramid of base edges 35 mm and vertical height 80 mm. Draw the given elevation and the true shape of the surface contained in the section plane X–X.

University of London School Examinations

28 Three views of a model of a steam turbine are shown in Figure 4.91. Draw, to a scale of half-full size, showing all hidden detail:

(a) the given plan and FE;
(b) an AP in the direction of 'KK'.
Associated Examining Board

29 Details of an angle bracket are shown in Figure 4.92. Draw the two given views, the view as seen from the direction of arrow *A* and an elevation

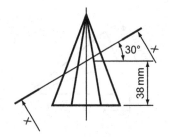

Figure 4.90

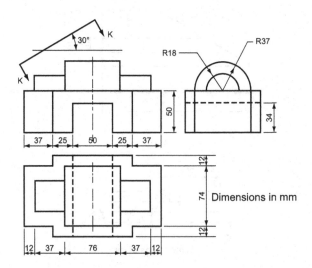

Figure 4.91

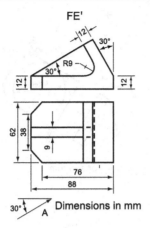

Figure 4.92

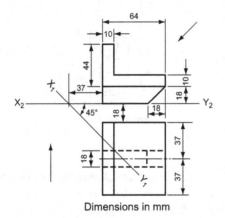

Figure 4.93

as seen from the left of the FE′. Hidden detail need not be shown. Scale: full size; use first angle orthographic projection.

Oxford Local Examinations

30 The plan and elevation of a special angle bracket are shown in Figure 4.93.

(a) Draw, full size, the given views and project an AP on the ground line X_1–Y_1.

(b) Using the AP in (a), project an AE on the ground line X_2–Y_2. All hidden detail to be shown.

Associated Examining Board

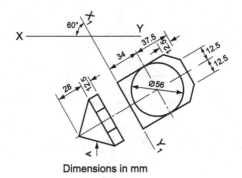

Dimensions in mm

Figure 4.94

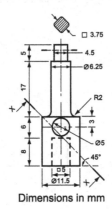

Dimensions in mm

Figure 4.95

31 Two views of a pivot block are shown in Figure 4.94. Draw the given views, and produce an elevation on XY as seen when looking in the direction of arrow A. Hidden edges are to be shown.

Cambridge Local Examinations

32 Figure 4.95 illustrates an elevation of an extension leg for a socket spanner. Draw, to a scale of 4:1 and in first angle orthographic projection:

(a) the given elevation;
(b) a plan;
(c) the true shape of the section at the cutting plane XX.
Oxford Local Examinations

Conic Sections

33 Figure 4.96 is the frustum of a right cone. Draw this elevation and a plan. Draw the true shape of the face AB.

Southern Regional Examinations Board

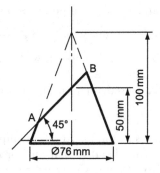

Figure 4.96

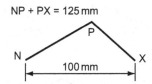

Figure 4.97

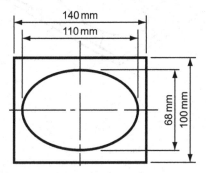

Figure 4.98

34 Figure 4.97 shows a point P which moves so that the sum of the distance from P to two fixed points, 100 mm apart, is constant and equal to 125 mm. Plot the path of the point P. Name the curve and the given fixed points.

Associated Lancashire Schools Examining Board

35 Figure 4.98 shows the loudspeaker grill of a car radio. The grill is rectangular with an elliptical hole. Draw the grill, full size, showing the construction of the ellipse clearly

West Midlands Examinations Board

36 Figure 4.99 shows an elliptical fish pond for a small garden. The ellipse is 1440 mm long and 720 mm wide. Using a scale of $\frac{1}{12}$, draw a true elliptical shape of the pond. (Do not draw the surrounding stones.) *All* construction must be shown.

    If a paper trammel is used, an accurate drawing of it must be made.

East Anglian Examinations Board

37 Figure 4.100 shows a section, based on an ellipse, for a handrail which requires cutting to form a bend so that the horizontal overall distance is increased from 112 to 125 mm. Construct the given figures and show the tangent construction at P and P_1. Show the true shape of the cut when the horizontal distance is increased from 112 to 125 mm.

    Southern Universities' Joint Board (See Chapter 5 for information not in Chapter 4.)

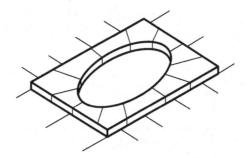

Figure 4.99

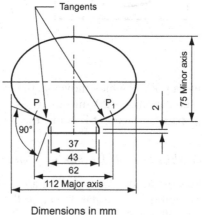

Dimensions in mm

Figure 4.100

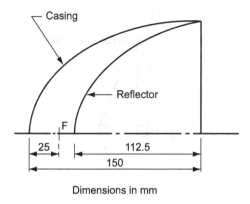

Dimensions in mm

Figure 4.101

38 Figure 4.101 shows the *upper half* of the section of a small headlamp.
The casing is in the form of a semi-ellipse. F is the focal point. The reflector section is in the form of a parabola.

Part 1. Draw, full size, the complete semi-ellipse.
Part 2. Draw, full size, the complete parabola inside the semi-ellipse.
Southern Regional Examinations Board

39 A point moves in a plane in such a way that its distance from a fixed
point is equal to its shortest distance from a fixed straight line. Plot the
locus of the moving point when the fixed point is 44 mm from the fixed
line. The maximum distance of the moving point is 125 mm from the
fixed point. State the name of the locus, the fixed point and the fixed line.

Associated Examining Board

40 A piece of wire is bent into the form of a parabola. It fits into a rectangle
which has a base length of 125 mm and a height of 100 mm. The open
ends of the wire are 125 mm apart. By means of a single line, show the
true shape of the wire.

Cambridge Local Examinations

41 An arch has a span of 40 m and a central rise of 13 m and the centre
line is an arc of a parabola. Draw the centre line of the arch to a scale of
10 mm = 20 m.

Oxford and Cambridge Schools Examination Board

42 A cone, vertical height 100 mm, base 75 mm diameter, is cut by a plane
parallel to its axis and 12 mm from it. Draw the necessary views to show
the true shape of the section and state the name of it.

Oxford and Cambridge Schools Examination Board

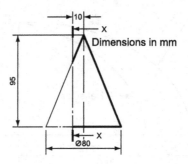

Figure 4.102

43 Figure 4.102 shows a right cone cut by a plane X–X. Draw the given view and project an elevation seen from the left of the given view.

44 Draw the conic having an eccentricity of $\frac{3}{4}$ and a focus which is 38 mm from the directrix. State the name of this curve.

Associated Examining Board

Chapter 5

Basic Concepts of Descriptive Geometry

In Chapter 4 the basic principles of orthographic projection have been addressed. The presented diagrams about multi-view projection should be thoroughly understood before this chapter is attempted. The object to be drawn is suspended between three planes called the FVP, the EVP and the horizontal plane HP. These planes are at right angles, and a view of the object is projected onto each of the planes. These views are called the FE, the EE and the plan. Two of the planes are then swung back, as if on hinges, until all three planes are in the same plane, i.e. they would all lie on the same flat surface. This system of swinging the planes until they are in line is called 'rabatment'.

Definitions

When a line passes through a plane, the point of intersection is called a 'trace'.

When a plane passes through another plane, the line of intersection is also called a 'trace'.

5.1 ORTHOGRAPHIC PROJECTION

5.1.1 Projection of Lines

The projection of a line that is not parallel to any of the principal planes.

Figure 5.1 shows a straight line AB suspended between the three principal planes. Projectors from A and B, perpendicular to the planes, give the projection of AB on each of the principal planes. On the right of Figure 5.1 can be seen the projections of the line after rabatment.

A trace is the line resulting from the intersection of two planes. The trace of the FVP and the HP is the line OX. The trace of the EVP and the HP is the line OY. The trace of the FVP and the EVP is the line OZ. These lines are often very useful for reference purposes and they should be marked on your drawings. The O is often ignored and the traces are then shown as XY and YZ.

To find the true length of a line that is not parallel to any of the principal planes and to find the angle that the line makes with the FVP (Figure 5.2).

The line is AB. On the FVP it is seen as ab and on the HP as a_1b_1.

DOI: 10.1201/9781003001386-5

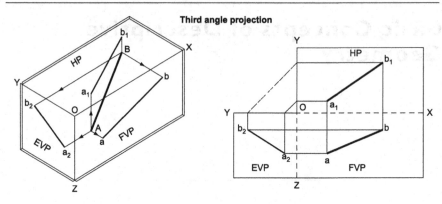

Figure 5.1 Projection of a line that is not parallel to any principal plane.

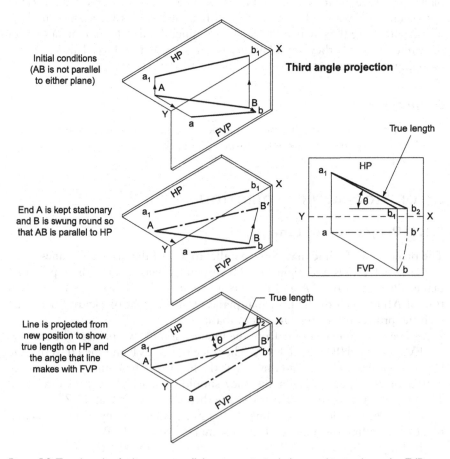

Figure 5.2 True length of a line not parallel to any principal plane and its angle to the FVP.

One end of the line A is kept stationary whilst B is swung round so that AB is parallel to the HP. B is now at B', and on the FVP, b is now at b'. Since the line is parallel to the HP, it will project its true length onto the HP. This is shown as a_1b_2. Note that b_1 and b_2 are the same distance from the line XY.

Since AB' (and ab') are parallel to the HP, the angle that AB makes with the FVP can be measured. This is shown as θ.

To find the true length of a line that is not parallel to any of the principal planes and to find the angle that the line makes with the HP (Figure 5.3).

The line is AB. On the FVP it is seen as ab and on the HP as a_1b_1.

One end of the line B is kept stationary while A is swung round so that AB is parallel to the FVP. A is now at A' and, on the HP, a is now at a'. Since the line is now parallel to the FVP it will project its true length onto the FVP. This is shown as a_2b. Note that a_2 and a are the same distance from the line XY.

Since BA (and b_1a') are parallel to the FVP, the angle that AB makes with the HP can be measured. This is shown as φ.

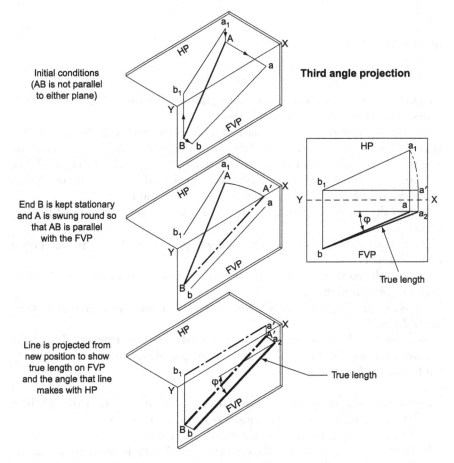

Figure 5.3 True length of a line not parallel to any principal plane and its angle to the HP.

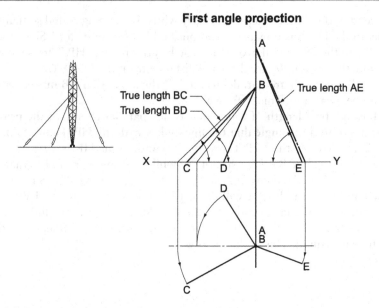

First angle projection

True length BC

True length BD

True length AE

Figure 5.4 Example of determining the true length of a line.

Figure 5.4 is an example of an application of the theory shown earlier. It shows how simple it is to apply this theory.

A pylon is supported by three hawsers. Given the plan and elevation of the hawsers, find their true lengths and the angle that they make with the ground.

In the plan, each hawser is swung round until it is parallel to the FVP. The new positions of the ends of the hawsers are projected up to the FE and joined to the pylon at A and B. This gives the true lengths and the angles.

To find the traces of a straight line given the plan and elevation of the line (Figure 5.5).

The line is AB. If the line is produced it will pass through both planes, giving traces T_v and T_h.

ab is produced to meet the XY line. This intersection is projected down to meet $a_1 b_1$ produced in T_h.

$b_1 a_1$ is produced to meet the XY line. This intersection is projected up to meet *ba* produced in T_v.

To draw the elevation and plan of a line AB given its true length and the distances of the ends of the line from the principal planes, in this case a_v **and** a_h, **and** b_v **and** b_h (Figure 5.6).

1 Fix points *a* and a_1 at the given distances a_v and a_h from the XY line. These are measured on a common perpendicular to XY.
2 Draw a line parallel to XY, distance b_v from XY.
3 With centre *a*, radius equal to the true length AB, draw an arc to cut the line drawn parallel to XY in C.

First angle projection

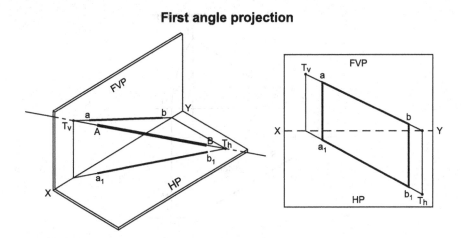

Figure 5.5 Finding the traces of a straight line.

First angle projection

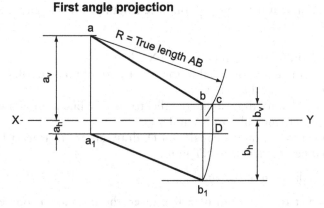

Figure 5.6 Elevation and plan of a line given its true length and distances from the principal planes.

4 From a_1 draw a line parallel to XY to meet a line from C drawn perpendicular to XY in

5 Draw a line parallel to XY, distance b_h from XY.

6 With centre a_1, radius a_1D, draw an arc to cut the line drawn parallel to XY in b_1.

7 Draw a line from b_1, perpendicular to XY to meet the line drawn parallel to XY through C in b.

ab is the elevation of the line.

a_1b_1 is the plan of the line.

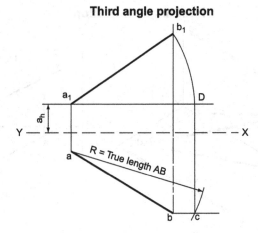

Figure 5.7 Plan of a line given a distance from the XY line, the true length and the elevation.

To construct the plan of a line AB given the distance of one end of the line from the XY line in the plan (a_h), the true length of the line and the elevation (Figure 5.7).

1   From b draw a line parallel to the XY line.
2   With centre a, radius equal to the true length of the line AB, draw an arc to cut the parallel line in C.
3   From a_1 (given), draw a line parallel to the XY line to meet a line drawn from C perpendicular to XY in D.
4   With centre a_1, radius equal to a_1D, draw an arc to meet a line drawn from b perpendicular to XY in b_1.

   $a_1 b_1$ is the plan of the line.

To construct the elevation of a line given the distance of one end of the line from the XY line in the elevation, the true length of the line and the plan (Figure 5.8).

This construction is very similar to the last one and can be followed from the instructions given for that example.

To construct the elevation of a line AB given the plan of the line and the angle that the line makes with the HP (Figure 5.9).

1   Draw the plan and from one end erect a perpendicular.
2   From the other end of the plan draw a line at the angle given to intersect the perpendicular in C.
3   From b_1 draw a line perpendicular to XY to meet XY in b.
4   From a_1 draw a line perpendicular to XY and mark off XY to a equal to a_1c.

   ab is the required elevation. An alternative solution is also shown.

Third angle projection

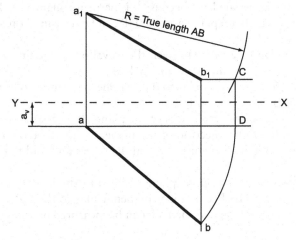

Figure 5.8 Elevation of a line given a distance from the XY line, the true length and the plan.

First angle projection

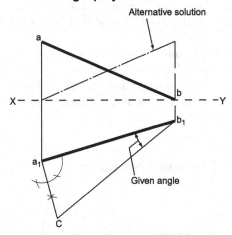

Figure 5.9 Elevation of a line given the plan and its angle with the HP.

5.1.2 Projection of Planes

The Inclined Plane

DEFINITION

An inclined plane is inclined to two of the principal planes and perpendicular to the third.

Figure 5.10 shows a rectangular plane that is inclined to the HP and the EVP and is perpendicular to the FVP. Because it is perpendicular to the FVP, the true angle between the inclined plane and the HP can be measured on the FVP. This is the angle φ.

To draw the plan of a line AB given the elevation of the line and the angle that the line makes with the VP (Figure 5.11).

This construction is very similar to the last one and can be followed from the instructions given for that example.

The top drawing shows the traces of the plane after rabatment. The bottom drawing shows the full projection of the plane. It should be obvious how the full projection is obtained if you are given the traces and told that the plane is rectangular.

Figure 5.12 shows a triangular plane inclined to the FVP and the EVP and perpendicular to the HP. Because it is perpendicular to the HP, the true angle between the inclined plane and the FVP can be measured on the HP. This angle is θ.

Once again, it should be obvious how the full projection of the inclined plane is obtained if you are given the traces and told that the plane is triangular.

To find the true shape of an inclined plane.

If the inclined plane is swung round so that it is parallel to one of the reference planes, the true shape can be projected. In Figure 5.13, the plan of the plane, HT, is swung round to H'T. The true shape of the plane can then be drawn on the FVP.

Figure 5.14 shows an example. An oblique, truncated, rectangular pyramid stands on its base. The problem is to find the true shape of sides A and B.

In the FE, side A is swung upright and its vertical height is projected across to the EE where the true shape of side A can be drawn.

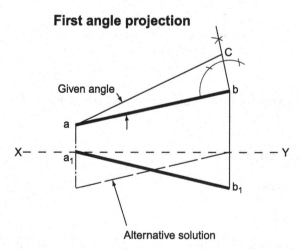

Figure 5.10 True angle of an inclined plane.

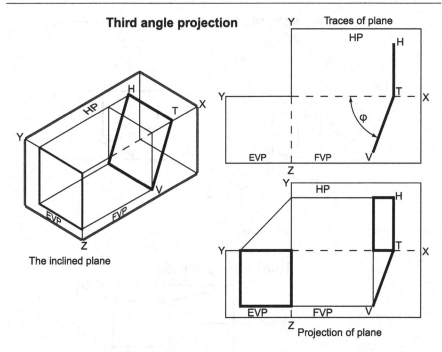

Figure 5.11 Plan of a line given the elevation and its angle with the VP.

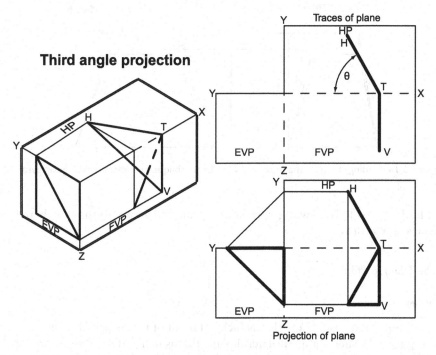

Figure 5.12 A triangular plane inclined to the FVP and the EVP and perpendicular to the HP.

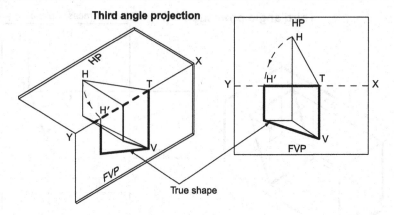

Figure 5.13 True shape of an inclined plane.

Third angle projection

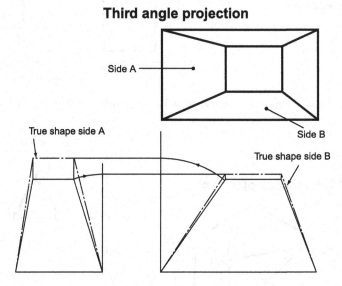

Figure 5.14 Finding the true shape of two sides of an oblique, truncated, rectangular pyramid.

In the EE, side B is swung upright and projected across to the FE where the true shape is drawn.

The Oblique Plane

DEFINITION

An oblique plane is a plane that is inclined to all of the principal planes.

Figure 5.15 shows a quadrilateral plane that is inclined to all three principal planes.

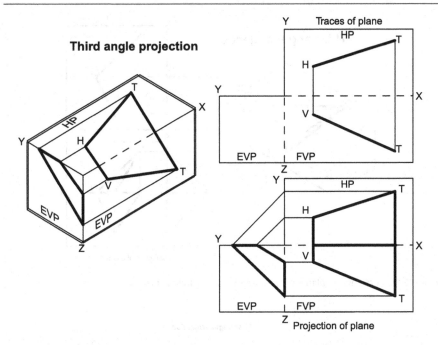

Figure 5.15 A quadrilateral plane inclined to all three principal planes.

The top drawing shows the traces of the plane after rabatment. The bottom drawing shows the full projection of the plane. It should be obvious how the projection is obtained if you are given the traces.

Figure 5.16 shows a triangular plane that is inclined to all three principal planes.

The top drawing shows the traces of the plane after rabatment. The bottom drawing shows the full projection of the plane. It should be obvious how the projection is obtained if you are given the traces.

To find the true angle between the HP and an oblique plane (Figure 5.17).

A triangle is inserted under the oblique plane and at right angles to it. This triangle, PQR, meets the HP at the same angle as the oblique plane.

The triangle is swung round until it is parallel to the FVP. Its new position is PQS and the angle required is PŜQ.

To find the true angle between the VP and an oblique plane (Figure 5.18).

A triangle is inserted under the oblique plane and at right angles to it. This triangle, PQR, meets the FVP at the same angle as the oblique plane.

The triangle is swung round until it is parallel to the HP. Its new position is PQS and the angle required is PŜQ.

To find the true angle between the traces of a given oblique plane VTH (Figure 5.19).

From any point *a* on the XY line draw *ab* (perpendicular to XY) and *ac* (perpendicular to TH).

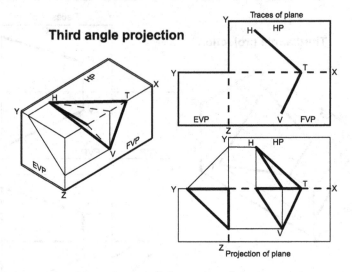

Figure 5.16 A triangular plane inclined to all three principal planes.

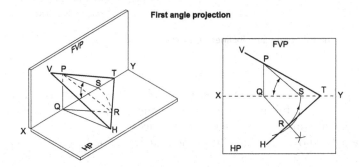

Figure 5.17 True angle between the HP and an oblique plane.

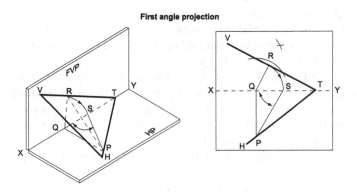

Figure 5.18 True angle between the VP and an oblique plane.

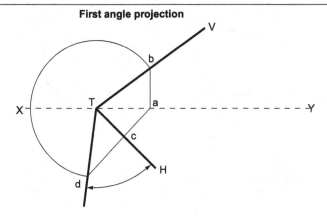

First angle projection

Figure 5.19 True angle between the traces of a given oblique plane.

With centre T, radius Tb, draw an arc to meet ac produced in d. $d\hat{T}c$ is the required angle.

5.2 ENLARGING AND REDUCING PLANE FIGURES AND EQUIVALENT AREAS

Definition

Similar figures are figures that have the same shape but may be different in size.

Constructions

To construct a figure, similar to another figure, having sides $\frac{7}{5}$ the length of the given figure.

Three examples, using the same basic method, are shown in Figure 5.20.

Select a point P, sometimes called the centre of similitude, in one of the positions shown.

From P draw lines through all the corners of the figure.

Extend the length of one of the lines from P to a corner, say PQ, in the ratio 7:5. The new length is PR.

Beginning at R, draw the sides of the larger figure parallel to the sides of the original smaller figure.

This construction works equally well for reducing the size of a plane figure.

Figure 5.21 shows an irregular hexagon reduced to $\frac{4}{9}$ its original size.

These constructions are practical only if the figure which has to be enlarged or reduced has straight sides. If the outline is irregular, a different approach is needed.

Figure 5.22 shows the face of a clown in two sizes, one twice that of the other. The change in size is determined by the two grids. A grid of known size

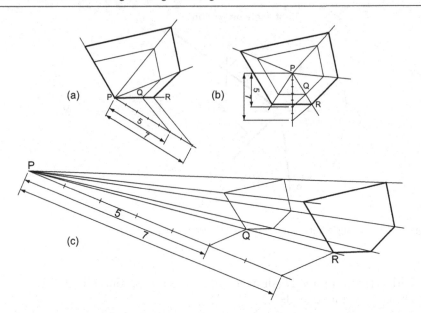

Figure 5.20 Constructing a similar enlarged fig. (a) using some of the original fig. (b) superimposing on original fig. (c) keeping outside the original figure.

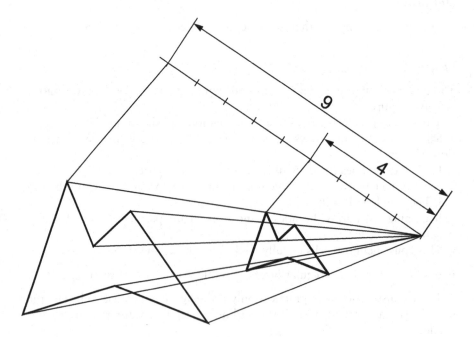

Figure 5.21 Reducing a figure outside the original figure.

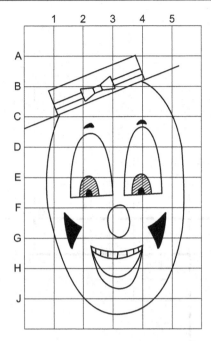

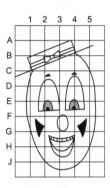

Figure 5.22 Enlarging and reducing an irregular shape.

is drawn over the first face and then another grid, similar to the first and at the required scale, is drawn alongside. Both grids are marked off, from A to J and from 1 to 5 in this case, and the points where the irregular outlines cross the lines of the grid are transferred from one grid to the other.

The closer together the lines of the grid, the greater the accuracy of the scaled copy.

It is sometimes necessary to enlarge or reduce a plane figure in one direction only. In this case, although the dimensions are changed, the proportions remain the same. Figure 5.23 shows a simple example of this. The figure has overall dimensions of 4 cm × 4 cm. The enlarged version retains the original proportions but now measures 6 cm × 4 cm.

First produce CA and BA. Mark off the new dimensions along CA and BA produced. This gives AB' and AC'.

Draw the square AB'XB and the rectangle ACYC' and draw the diagonals AX and AY.

From points along the periphery of the original plane figure (in this case 1 to 10), draw lines horizontally and vertically to and from the diagonals to intersect in 1', 2', 3', etc. Points 1' to 10' give the new profile.

The enlarged or reduced figures produced in Figure 5.23 are mirror images of the original figures. Usually this does not matter, particularly if the figure is for a template; it just has to be turned over. However, if it does matter, a

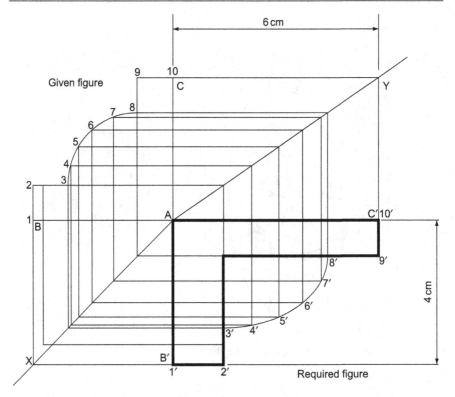

Figure 5.23 Enlarging or reducing a place figure in one direction only.

construction similar to that used in Figure 5.24 must be used. In this case, a basic 60 mm × 40 mm shape has been changed into a 30 mm × 20 mm shape.

A′B′ and A′C′ are drawn parallel to AB and AC and marked off 20 and 30 mm long, respectively.

AA′ and BB′, AA′ and CC′ are produced to meet in Q and P, respectively.

The curved part of the figure is divided into as many parts as is necessary to produce an accurate copy.

The rest of the construction should be self-explanatory.

The transfer of the markings along A′B′ and A′C′ on the original figure to the required figure is made easier by the use of a 'trammel'; this is a rather pomp-ous title for a piece of paper with a straight edge. If you lay this piece of paper along AC on the given figure and mark off A, C and all the relevant points in between, you can line up the paper with A and C on the required figure and transfer the points between A′ and C′ onto the required figure. The same thing can be done for A′B′.

All the changes of shape so far have been dependent upon a known change of length of one or more of the sides. No consideration has been made of a spe-cific change of area. The ability to enlarge or reduce a given shape in terms of area has applications. If, for instance, fluid flowing in a pipe is divided into two

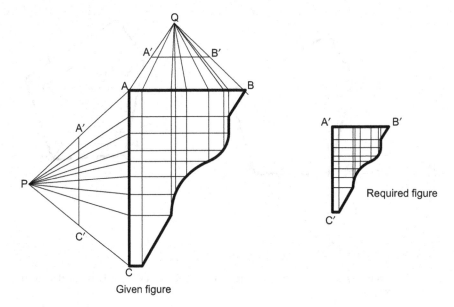

Given figure

Required figure

Figure 5.24 Reducing a plane figure in two directions.

smaller pipes of equal area, then the area of the larger pipe will be twice that of the two smaller ones. This does not mean, of course, that the dimensions of the larger pipe are twice that of the smaller ones.

Select a point P. (This may be on a corner, or within the outline of the pentagon, or outside the outline although this is not shown because the construction is very large.)

Let A be a corner of the given pentagon.

Join PA and produce it.

Draw a semi-circle, centre P, radius PA.

From P, drop a perpendicular to PA to meet the semi-circle in S.

Mark off PR:PQ in the required ratio, in this case 2:1.

Bisect AR in O, and erect a semi-circle, radius OR to cut PS produced in T.

Join SA and draw TA' parallel to SA.

A' is the first corner of the enlarged pentagon.

Although Figure 5.25 shows a pentagon, the construction applies to any plane figure and can be used to increase and decrease a plane figure in a known ratio of areas.

Equivalent Areas

To construct a rectangle equal in area to a given triangle ABC (Figure 5.26).

1   From B, the apex of the triangle, drop a perpendicular to meet the base in F.

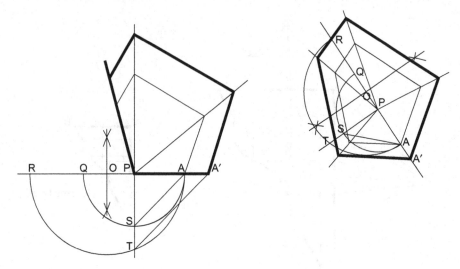

Figure 5.25 Alternatives for enlarging a pentagon so that its new area is twice that of the original.

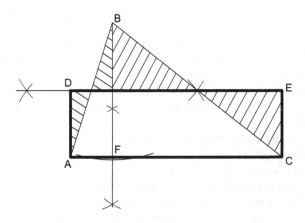

Figure 5.26 To construct a rectangle equal in area to a given triangle.

2   Bisect FB.
3   From A and C erect perpendiculars to meet the bisected line in D and E.

ADEC is the required rectangle.

It should be obvious from the shading that the part of the triangle that is outside the rectangle is equal in area to that part of the rectangle that overlaps the triangle.

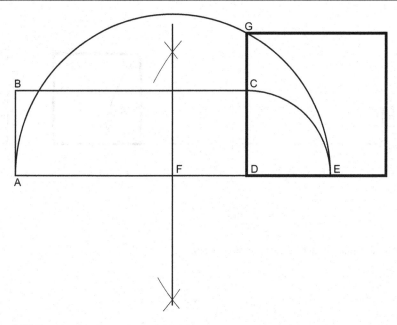

Figure 5.27 To construct a square equal in area to a given rectangle.

To construct a square equal in area to a given rectangle ABCD (Figure 5.27).

1 With centre D, radius DC, draw an arc to meet AD produced in E.
2 Bisect AE and erect a semi-circle, radius AF, centre F.
3 Produce DC to meet the semi-circle in G.

DG is one side of the square. (For the construction of a square, given one of the sides, see Chapter 3.)

This construction can be adapted to find the square root of a number. Figure 5.28 shows how to find $\sqrt{6}$.

Since the area of the rectangle equals that of the square, then

$$ab = c^2$$

If *a* is always = 1, then $b = c^2$
or

$$\sqrt{b} = c$$

Thus, always draw the original rectangle with one side equal to one unit, and convert the rectangle into a square of equal area.

To construct a square equal in area to a given triangle (Figure 5.29).

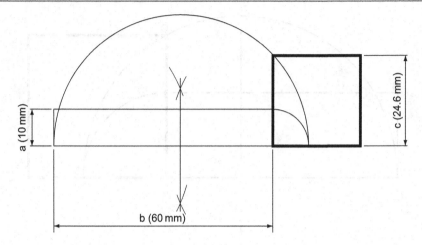

Figure 5.28 Finding a square route geometrically.

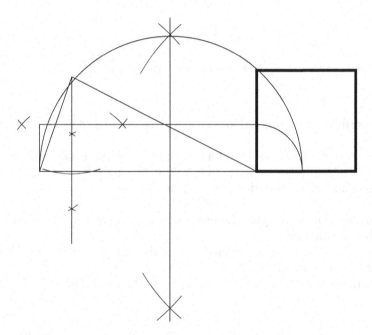

Figure 5.29 To construct a square equal in area to a given triangle.

This construction is a combination of those described in Figure 5.26 and Figure 5.27. First change the triangle into a rectangle of equivalent area and then change the rectangle into a square of equivalent area.

To construct a triangle equal in area to a given polygon ABCDE (Figure 5.30).

1 Join CE and from D draw a line parallel to CE to meet AE produced in F.
2 Join CF.

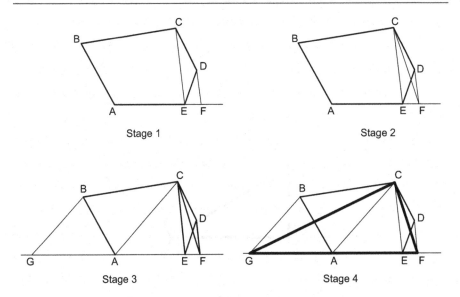

Figure 5.30 To construct a triangle equal in area to a given polygon.

3   Since DF is parallel to CE, triangles CDE and CFE have the same base and vertical height and therefore the same area. The polygon ABCDE now has the same area as the quadrilateral ABCF and the original five-sided figure has been reduced to a four-sided figure of the same area.

4   Join CA and from B draw a line parallel to CA to meet EA produced in G.

5   Join CG.

6   Since BG is parallel to CA, triangles CBA and CGA have the same base and vertical height and therefore the same area. The quadrilateral ABCF now has the same area as the triangle GCF and the original five-sided figure has been reduced to a three-sided figure of the same area.

GCF is the required triangle.

The theorem of Pythagoras says that 'in a right-angled triangle, the square on the hypotenuse is equal to the sum of the squares on the other two sides'. When this theorem is shown pictorially, it is usually illustrated by a triangle with squares drawn on the sides. This tends to be a little misleading since the theorem is valid for any similar plane figures (Figure 5.31).

This construction is particularly useful when you wish to find the size of a circle that has the equivalent area of two or more smaller circles added together.

To find the diameter of a circle that has the same area as two circles, 30 mm and 50 mm diameter (Figure 5.32).

Draw a line 30 mm long.

From one end erect a perpendicular, 50 mm long.

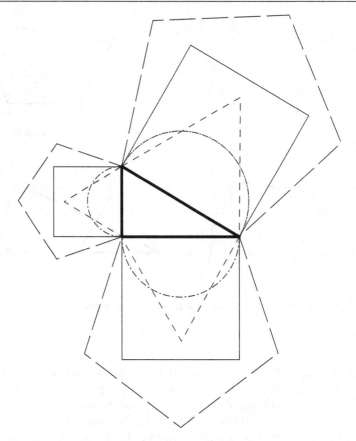

Figure 5.31 An illustration of the theorem of Pythagoras.

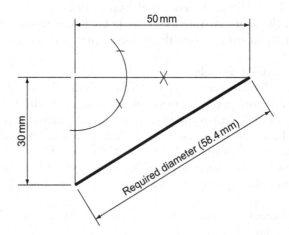

Figure 5.32 Using Pythagoras to find equivalent areas.

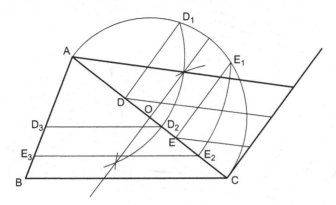

Figure 5.33 To divide a triangle into three parts of equal area.

The hypotenuse of the triangle thus formed is the required diameter (58.4 mm).

If you have to find the single equivalent diameter of more than two circles, reduce them in pairs until you have two, and then finally one left.

To divide a triangle ABC into three parts of equal area by drawing lines parallel to one of the sides (e.g. BC) (Figure 5.33).

1   Bisect AC (or AB) in O and erect a semi-circle, centre O, radius OA.
2   Divide AC into three equal parts AD, DE and EC and erect perpendiculars from D and E to meet the semi-circle in D_1 and E_1.
3   With centre A and radius AD_1, draw an arc to cut AC in D_2.
4   With centre A and radius AE_1, draw an arc to cut AC in E_2.
5   From D_2 and E_2, draw lines parallel to BC to meet AB in D_3 and E_3, respectively.

The areas AD_2D_3, $D_3D_2E_2E_3$ and E_3E_2CB are equal.

Although Figure 5.33 shows a triangle divided into three equal areas, the construction can be used for any number of equal areas.

To divide a polygon into a number of equal areas (e.g. five) by lines drawn parallel to the sides.

This construction is very similar to that used for Figure 5.33. Proceed as for the triangle and complete as shown in Figure 5.34.

Again, this construction can be used for any polygon and can be adapted to divide any polygon into any number of equal areas.

5.3 INTERSECTION OF REGULAR SOLIDS

When two solids interpenetrate, a line of intersection is formed. It is sometimes necessary to know the exact shape of this line, usually so that an accurate development of either or both of the solids can be drawn. This section shows

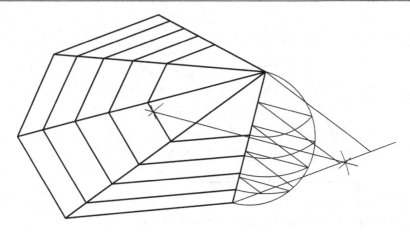

Figure 5.34 To divide a polygon into a number of equal areas.

the lines of intersection formed when some of the simpler geometric solids interpenetrate.

Two dissimilar square prisms meeting at right angles (Figure 5.35).

The EE shows where corners 1 and 3 meet the larger prism, and these are projected across to the FE. The plan shows where corners 2 and 4 meet the larger prism, and this is projected up to the FE.

Two dissimilar square prisms meeting at an angle (Figure 5.36).

The FE shows where corners 1 and 3 meet the larger prism. The plan shows where corners 2 and 4 meet the larger prism and this is projected down to the FE.

A hexagonal prism meeting a square prism at right angles (Figure 5.37).

The plan shows where all the corners of the hexagonal prism meet the square prism. These are projected down to the FE to meet the projectors from the same corners on the EE.

Two dissimilar hexagonal prisms meeting at an angle (Figure 5.38).

The FE shows where corners 3 and 6 meet the larger prism. The plan shows where corners 1, 2, 4 and 5 meet the larger prism and these are projected up to the FE.

A hexagonal prism meeting an octagonal prism at an angle, their centres not being in the same VP (Figure 5.39).

The FE shows where corners 3 and 6 meet the octagonal prism. The plan shows where corners 1, 2, 4 and 5 meet the octagonal prism and these are projected down to the FE.

The sides of the hexagonal prism between corners 3–4 and 5–6 meet two sides of the octagonal prism. The change of shape occurs at points *a* and *b*. The position of *a* and *b* on the FE (and then across to the EE) is found by projecting down to the FE via the end of the hexagonal prism (follow the arrows). The intersection on the FE can then be completed.

First angle projection

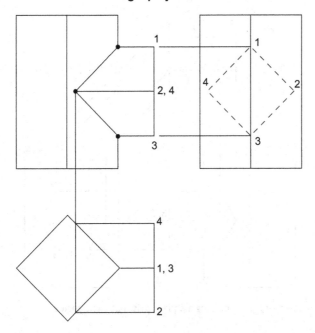

Figure 5.35 Two dissimilar square prisms meeting at right angles.

Third angle projection

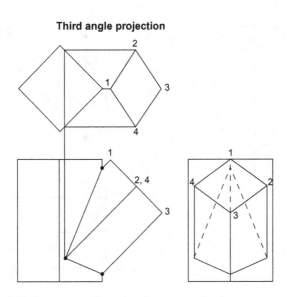

Figure 5.36 Two dissimilar square prisms meeting at an angle.

Third angle projection

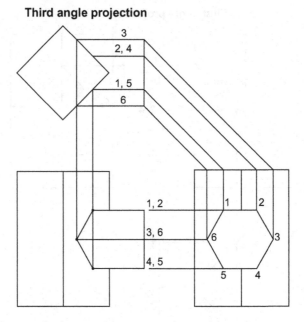

Figure 5.37 A hexagonal prism meeting a square prism at right angles.

First angle projection

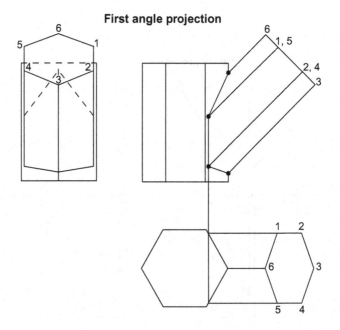

Figure 5.38 Two dissimilar hexagonal prisms meeting at an angle.

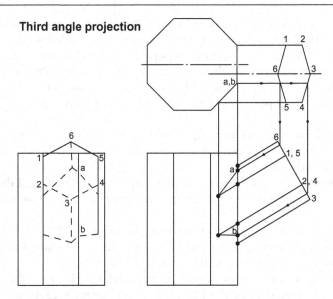

Figure 5.39 A hexagonal prism meeting an octagonal prism at an angle.

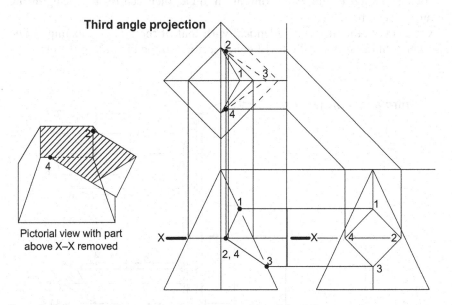

Figure 5.40 A square prism meeting a square pyramid at right angles.

A square prism meeting a square pyramid at right angles (Figure 5.40).
The EE shows where corners 1 and 3 meet the pyramid.
These are projected across to the FE.
Corners 2 and 4 are not quite so obvious. The pictorial view shows how
these corners meet the pyramid. If the pyramid was cut across X–X, the section

of the pyramid resulting would be square, and points 2 and 4 would lie on this square. It is not necessary to make a complete, shaded section on your drawing but it is necessary to draw the square on the plan. Since points 2 and 4 lie on this square, it is simple to find their exact position. Project corners 2 and 4 from the EE onto the plan. The points where these projectors meet the square are the exact positions of the intersections of corners 2 and 4 with the pyramid.

Two dissimilar cylinders meeting at right angles (Figure 5.41).

The smaller cylinder is divided into 12 equal sectors on the FE and on the plan (the EE shows how these are arranged round the cylinder).

The plan shows where these sectors meet the larger cylinder and these intersections are projected down to the FE to meet their corresponding sector at 1', 2', 3', etc.

Two dissimilar cylinders meeting at an angle (Figure 5.42).

The method is identical with that of the last problem. The smaller cylinder is divided into 12 equal sectors on the FE and on the plan.

The plan shows where these sectors meet the larger cylinder and these intersections are projected up to the FE to meet their corresponding sector at 1', 2', 3' etc.

Two dissimilar cylinders meeting at an angle, their centres not being in the same VP (Figure 5.43).

Once again, the method is identical with that of the previous example. The smaller cylinder is divided into 12 equal sectors on the FE and on the plan.

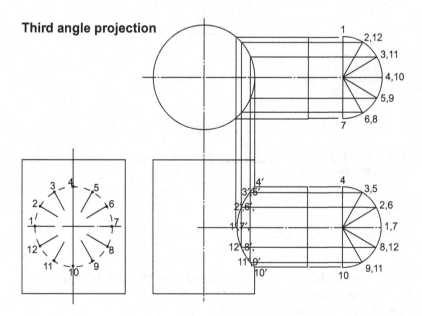

Figure 5.41 Two dissimilar cylinders meeting at right angles.

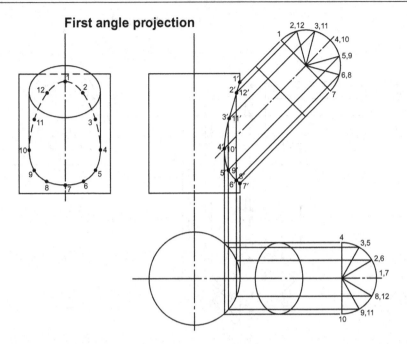

Figure 5.42 Two dissimilar cylinders meeting at an angle.

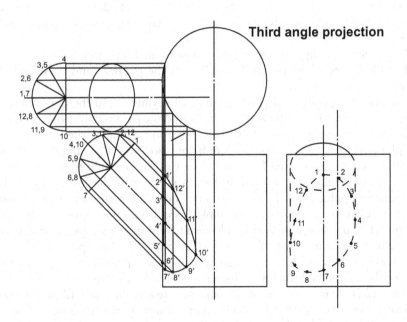

Figure 5.43 Two dissimilar cylinders meeting at an angle, their centres not being in the same VP.

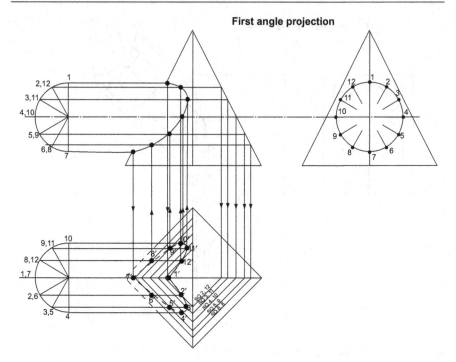

Figure 5.44 A cylinder meeting a square pyramid at right angles.

The plan shows where the sectors meet the larger cylinder and these intersections are projected down to the FE to meet their corresponding sectors at 1′, 2′, 3′, etc.

A cylinder meeting a square pyramid at right angles (Figure 5.44).

The FE shows where points 1 and 7 meet the pyramid and these are projected down to the plan.

Consider the position of point 2. Since the cylinder and the pyramid interpenetrate, point 2 lies on both the cylinder and the pyramid. Its position on the cylinder is seen fairly easily. On the FE it lies on the line marked 2,12 and on the plan it lies on the line marked 2,6. Its position on the pyramid is not quite so obvious. Imagine on the FE that the part of the pyramid that is above the line 2,12 was removed. The section that resulted across the pyramid would be a square and point 2 would lie somewhere along the perimeter of that square. It is not necessary to construct a complete, shaded section across the pyramid at line 2,12 but the square that would result from such a section is constructed on the plan. In Figure 5.44 this is marked as 'SQ 2,12'. Since point 2 lies somewhere along the line 2,6 (in the plan), then its exact position is at the intersection of the square and the line. This is shown on the plan as 2′.

Point 12 is the intersection of the same square and the line 8,12 (in the plan).

This process is repeated for each point in turn. When the intersection has been completed in the plan, it is a simple matter to project the points up onto the FE and draw the intersection there.

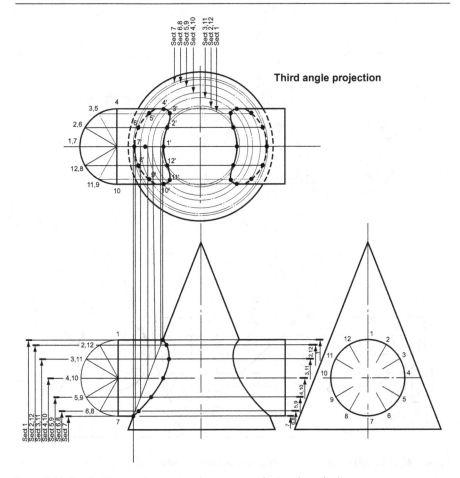

Figure 5.45 A cylinder meeting a cone, the cone enveloping the cylinder.

A cylinder meeting a cone, the cone enveloping the cylinder (Figure 5.45).

The cylinder is divided into 12 equal sectors on the FE and on the plan.

Consider point 2. On the FE it lies somewhere along the line marked 2,12 whilst on the plan it lies on the line marked 2,6. If, on the FE, that part of the cone above the line 2,12 was removed, point 2 would lie somewhere on the perimeter of the resulting section. In this case, the section of the cone is a circle and the radius of that circle is easily projected up to the plan. In Figure 5.45, the section is marked on the plan as 'Sect 2,12' and the exact position of point 2′ is the intersection of that section and the line marked 2,6. Point 12 is the intersection of the same section and the line marked 12,8.

This process is repeated for each point in turn. When the plan is complete, the points can be projected down to the FE; this is not shown for clarity.

A cylinder and a cone, neither enveloping the other (Figure 5.46).

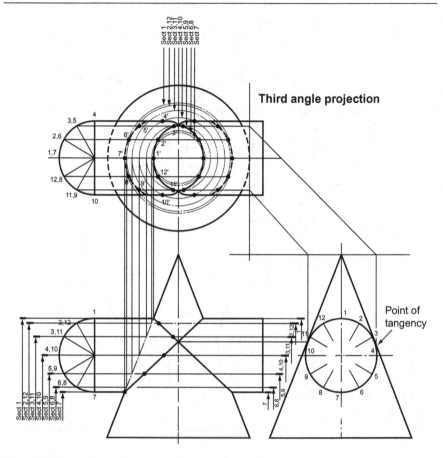

Figure 5.46 A cylinder and a cone, neither enveloping the other.

The constructions are exactly the same as those used in the previous example with one small addition.

The EE shows a point of tangency between the cylinder and the cone. This point is projected across to the FE and up to the plan as shown.

A cylinder and a cone, the cylinder enveloping the cone (Figure 5.47).

The construction required here is a modified version of the two previous ones. Instead of the cylinder being divided into 12 equal sectors, some of which would not be used, several points are selected on the EE. These are marked on the top part of the cylinder as *a*, *b* and *c* whilst the lower part is marked 1, 2, 3 and 4.

As before, the sections of the cone across each of these points are projected up to the plan from the FE. Each point is then projected from the EE to meet its corresponding section on the plan at *a'*, *b'*, *c'*, 1', 2', 3' and 4'.

These points are then projected down to the FE. For the sake of clarity, this is not shown.

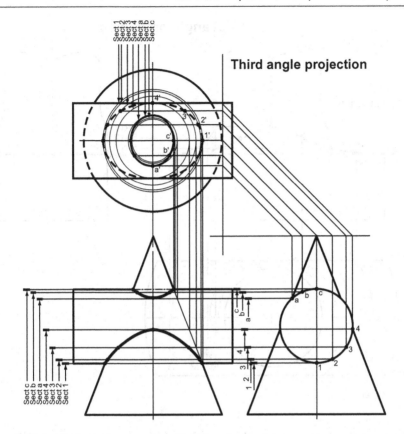

Figure 5.47 A cylinder and a cone, the cylinder enveloping the cone.

Fillet Curves

A sudden change of shape in any load-bearing component produces a stress centre, that is, an area that is more highly stressed than the rest of the component and therefore more liable to fracture under load. To avoid these sharp corners, fillet radii are used. These radii allow the stress to be distributed more evenly, making the component stronger.

Sometimes, parts of these fillet radii are removed and a curve of intersection results. Figure 5.48 shows an example of this.

Sections are taken on the FE. These appear on the plan as circles. The points where these sections 'run off' the plan can easily be seen (at 1, 2, 3 and 4) and they are projected up to the FE to meet their respective sections in 1', 2', 3' and 4'.

5.4 DEVELOPMENTS

There are three basic ways of fashioning a piece of material into a given shape. Either you start with a solid lump and take pieces off, you can cast or mould an appropriate material or you can bend sheet material as required.

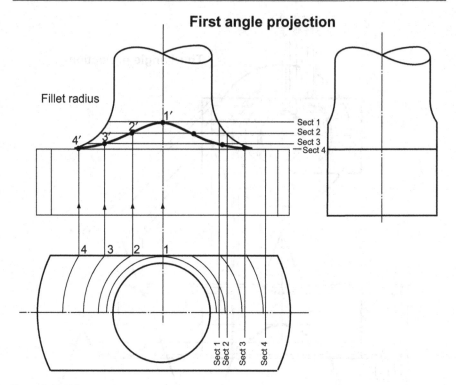

Figure 5.48 Fillet curves.

It should be obvious that, if the last method is used, the sheet material must first be shaped so that, after it is bent, you have the correct size and shape. If, then, a component is to be made of sheet material, the designer must not only visualise and draw the final three-dimensional component, but also calculate and draw the shape of the component in the form that it will take when marked out on the two-dimensional sheet material.

The process of unfolding the three-dimensional 'solid' is called 'development'.

The shapes of most engineering components made of sheet material are whole, or parts of, prisms, pyramids, cylinders or cones, so this section deals with the development of the shapes.

5.4.1 Prisms

Figure 5.49 shows how a square prism is unfolded and its development obtained.

Note that where there are corners in the undeveloped solid, these are shown as dotted lines in the development.

To develop a hexagonal prism with oblique ends (Figure 5.50).

The height of each corner of the development is found by projecting directly from the orthographic view.

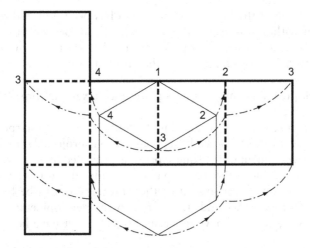

Figure 5.49 Developing a square prism.

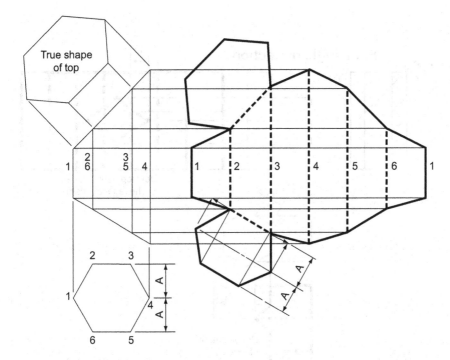

Figure 5.50 Developing a hexagonal prism.

The shapes of the top and the bottom are found by projecting the true shapes of the oblique faces. The top has been found by conventional means. The true shape is projected from the elevation and transferred to the development.

The true shape of the bottom of the prism has been drawn directly on the development without projecting the true shape from the elevation. The corner between lines 2 and 3 has been produced until it meets the projectors from corners 1 and 4. The produced line is then turned through 90° and the width, 2A, marked on.

The development of intersecting square and hexagonal prisms meeting at right angles (Figure 5.51).

First an orthographic drawing is made and the line of interpenetration is plotted. The development of the hexagonal prism is projected directly from the FE and the development of the square prism is projected directly from the plan.

Projecting from the orthographic views provides much of the information required to develop the prisms; any other information can be found on one of the orthographic views and transferred to the developments. In this case, dimensions A, *b*, C and *d* have not been projected but have been transferred with dividers.

The development of intersecting hexagonal and octagonal prisms meeting at an angle (Figure 5.52).

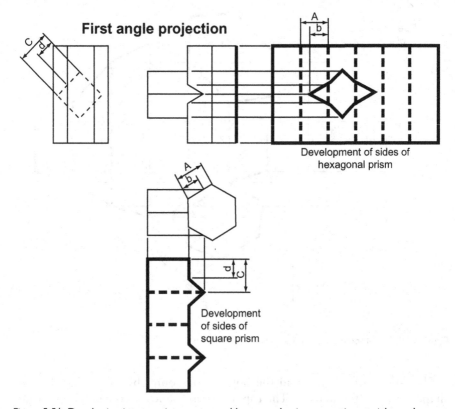

Figure 5.51 Developing intersecting square and hexagonal prisms meeting at right angles.

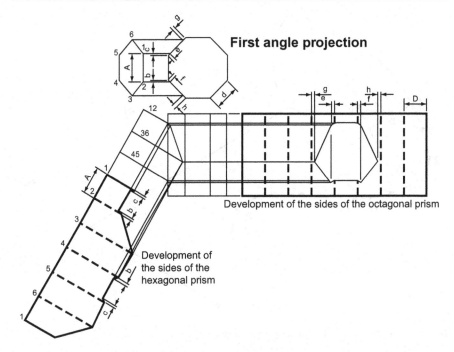

First angle projection

Development of the sides of the octagonal prism

Development of
the sides of the
hexagonal prism

Figure 5.52 Developing intersecting hexagonal and octagonal prisms meeting at an angle.

The method of developing these prisms is identical to that used in the previous example. This example is more complicated but the developments are still projected from one of the orthographic views, and any information that is not projected across can be found on the orthographic views and transferred to the development. In this case, dimensions A, *b*, *c*, D, etc., have not been projected but have been transferred with dividers.

5.4.2 Cylinders

If you painted the curved surface of a cylinder and, while the paint was wet, placed the cylinder on a flat surface and then rolled it once, the pattern that the paint left on the flat surface would be the development of the curved surface of the cylinder. Figure 5.53 shows the shape that would evolve if the cylinder were cut obliquely at one end. The length of the development would be πD, the circumference.

The oblique face has been divided into 12 equal parts and numbered. You can see where each number will touch the flat surface as the cylinder is rolled.

Figure 5.54 shows how the previous idea is interpreted into an accurate development of a cylinder.

To develop a cylinder with an oblique top (Figure 5.54).

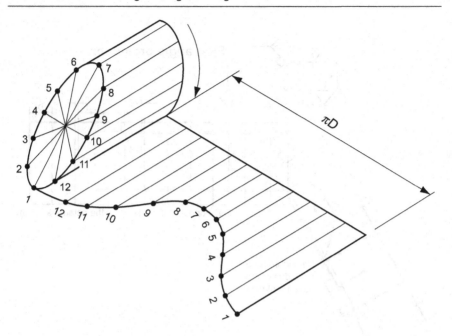

Figure 5.53 Pictorial view of the development of a cylinder.

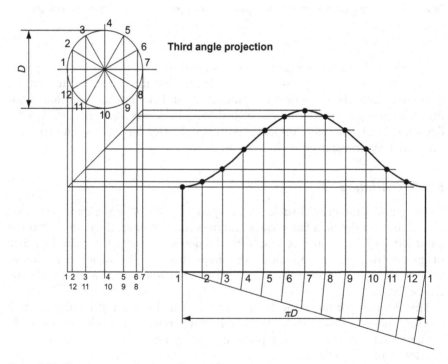

Figure 5.54 Developing a cylinder with an oblique top.

A plan and elevation of the cylinder is drawn. The plan is divided into 12 equal sectors which are numbered. These numbers are also marked on the elevation.

The circumference of the cylinder is calculated and is marked out alongside the elevation. This circumference πD is divided into 12 equal parts and these parts are numbered 1 to 12 to correspond with the 12 equal sectors.

The height of the cylinder at sector 1 is projected across to the development and a line is drawn up from point 1 on the development to meet the projector.

The height of the cylinder at sectors 2 and 12 is projected across to the development and lines are drawn up from points 2 and 12 on the development to meet the projector.

This process is repeated for all 12 points and the intersections are joined with a neat curve.

5.4.3 Pyramids

Figure 5.55 shows how the development of a pyramid is found. If a pyramid is tipped over so that it lies on one of its sides and is then rolled so that each of its sides touches in turn, the development is traced out. The development is formed within a circle whose radius is equal to the true length of one of the corners of the pyramid.

To develop the sides of the frustum of a square pyramid (Figure 5.56).

The true length of a corner of the pyramid can be seen in the FE. An arc is drawn, radius equal to this true length, centre the apex of the pyramid. A second arc is drawn, radius equal to the distance from the apex of the cone to the beginning of the frustum, centre the apex of the cone. The width of one side of the pyramid, measured at the base, is measured on the plan and this is stepped round the larger arc four times.

To develop the sides of a hexagonal frustum if the top has been cut obliquely (Figure 5.57).

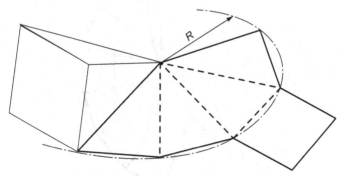

R = True length of a corner of the pyramid

Figure 5.55 Developing a pyramid.

First angle projection

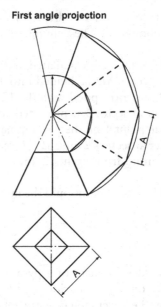

Figure 5.56 Developing the sides of the frustum of a square pyramid.

Third angle projection

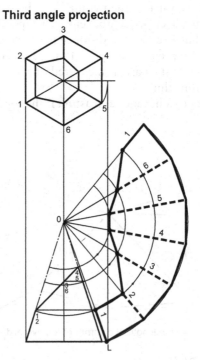

Figure 5.57 Developing the sides of a hexagonal frustum if the top has been cut obliquely.

The FE does not show the true length of a corner of the pyramid. Therefore, the true length, OL, is constructed and an arc, radius OL and centre O, is drawn. The width of one side of the pyramid, measured at the base, is stepped around the arc six times and the six sides of the pyramid are marked on the development.

The FE does not show the true length of a corner of the pyramid; equally, it does not show the true distance from O to any of the corners 1 to 6. However, if each of these corners is projected horizontally to the line OL (the true length of a corner), these true distances will be seen. With compass centre at O, these distances are swung round to their appropriate corners.

5.4.4 Cones

Figure 5.58 shows how if a cone is tipped over and then rolled it will trace out its development. The development forms a sector of a circle whose radius is equal to the slant height of the cone. The length of the arc of the sector is equal to the circumference of the base of the cone.

If the base of the cone is divided into 12 equal sectors that are numbered from 1 to 12, the points where the numbers touch the flat surface as the cone is rolled can be seen.

To develop the frustum of a cone.

The plan and elevation of the cone are shown in Figure 5.59. The plan is divided into 12 equal sectors. The arc shown as dimension A is $\frac{1}{12}$ of the circumference of the base cone.

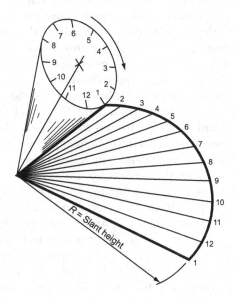

Figure 5.58 Development of a cone.

First angle projection

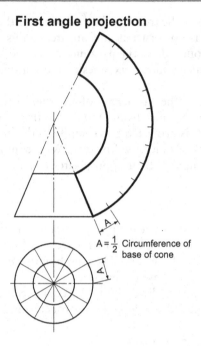

$A = \frac{1}{2}$ Circumference of base of cone

Figure 5.59 To develop the frustum of a cone.

With centre at the apex of the cone draw two arcs, one with a radius equal to the distance from the apex to the top of the frustum (measured along the side of the cone) and the other equal to the slant height of the cone.

With dividers measure distance A and step this dimension around the larger arc 12 times. (This will not give an exact measurement of the circumference at the base of the cone but it is a good approximation.)

To develop the frustum of a cone that has been cut obliquely (Figure 5.60).

Divide the plan into 12 equal sectors and number them from 1 to 12. Project these down to the FE and draw lines from each number to the apex A. You can see where each of these lines crosses the oblique top of the frustum. Now draw the basic development of the cone and number each sector from 1 to 12 and draw a line between each number and the apex A.

The lines A_1 and A_7 on the FE are the true length of the slant height of the cone. In fact, all of the lines from A to each number are equal in length but, on the FE, lines A_2 to A_6 and A_8 to A_{12} are shorter than A_1 and A_7 because they are sloping 'inwards' towards A. The true lengths from A to the oblique top of the frustum on these lines are found by projecting horizontally across to the line A_1. Here, the true length can be swung round with compasses to its respective sector and the resulting series of points joined together with a neat curve.

Third angle projection

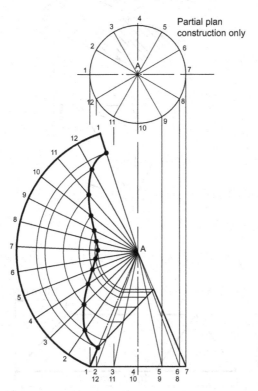

Partial plan
construction only

Figure 5.60 To develop the frustum of a cone that has been cut obliquely.

5.5 PROBLEMS

(All questions originally set in imperial units.)

Projection of Lines and Planes

1 Figure 5.61 shows the shaped end of a square fence post. Lengths AD, BD and DC are equal.

   By means of a suitable geometric construction, find (a) the true length of BD, (b) the true shape of the face ABD.
   Associated Lancashire Schools Examining Board

2 Figure 5.62 gives the FE and plan of the roof of a house in first angle projection. Draw the given views to a suitable scale and from them find, by construction, the true lengths of the rafters A and B. Print these lengths under your drawing.

   East Anglian Regional Examinations Board

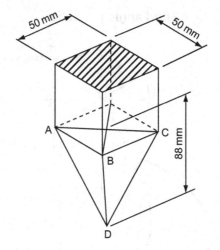

Figure 5.61

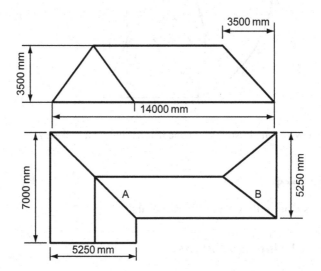

Figure 5.62

3 Figure 5.63 shows the projected elevation of a line AB which has a true length of 100 mm. End B of the line is 12 mm in front of the VP; end A is also in front of the VP. Draw the plan a, b and elevation of this line and determine and indicate its VT and HT. Measure, state and indicate the angle of inclination of the line to the HP.

Joint Matriculation Board

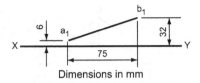

Dimensions in mm

Figure 5.63

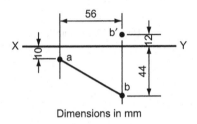

Dimensions in mm

Figure 5.64

4 A line AB of true length 88 mm lies in an auxiliary VP which makes an angle of 30° with the VP. The line is inclined at an angle of 45° to the horizontal, the point B being the lowest at a vertical distance of 12 mm above the HP and 12 mm in front of the VP. Draw the plan and elevation of AB and clearly identify them in the drawing.

Associated Examining Board

5 The plan of a line 82 mm long is shown in Figure 5.64. The elevation of one end is at *b'*. Complete the elevation and measure the inclinations of the line to the HP and VP.

University of London School Examinations

6 The projections of a triangle RST are shown in Figure 5.65. Determine the true shape of the triangle.

Associated Examining Board

7 The plan and elevation of two straight lines are given in Figure 5.66. Find the true lengths of the lines, the true angle between them and the distance between A and C.

Southern Universities' Joint Board

8 Figure 5.67 shows two views of an oblique triangular pyramid, standing on its base. Draw the given view together with an auxiliary view looking in the direction of arrow A which is perpendicular to BC. Also draw the

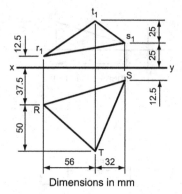

Dimensions in mm

Figure 5.65

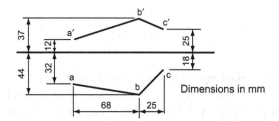

Dimensions in mm

Figure 5.66

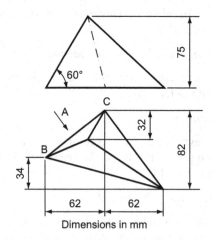

Dimensions in mm

Figure 5.67

true shapes of the sides of the pyramid. Note, omit dimensions but show hidden detail in all views. Scale: full size.

Oxford Local Examinations

Enlarging and Reducing Plane Figures and Equivalent Areas

9

    (a) Construct the triangle ABC, shown in Figure 5.68, from the information given and then construct a triangle CDA in which CD and CB are in the ratio 5:7.

    (b) Construct the polygon ABCDE shown in Figure 5.69. Construct by the use of radiating lines a polygon PQRST similar to ABCDE so that both polygons stand on the line PT.

Southern Regional Examinations Board

10 Figure 5.70 shows a sail for a model boat. Draw the figure, full size, and construct a similar shape with the side corresponding to AB 67 mm long.

Middlesex Regional Examining Board

11 Without the use of a protractor or set squares, construct a polygon ABCDE standing on a base AB given that AB = 95 mm, BC = 75 mm, CD = 55 mm, AE = 67.5 mm, ∠ABC = 120°, ∠EAB = 82½° and ∠CDE = 90°. Also construct a similar but larger polygon so that the

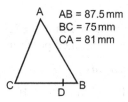

AB = 87.5 mm
BC = 75 mm
CA = 81 mm

Figure 5.68

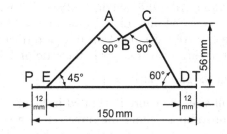

Figure 5.69

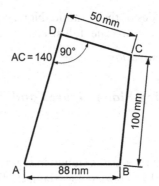

Figure 5.70

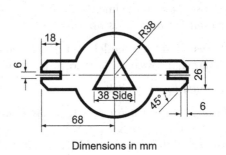

Dimensions in mm

Figure 5.71

side corresponding with AB becomes 117.5 mm. Measure and state the lengths of the sides of the enlarged polygon.

Oxford Local Examinations

12 In the triangle ABC, AB = 82 mm, BC = 105 mm and CA = 68 mm. Draw a triangle similar to ABC but having an area one-fifth the area of ABC.

Oxford and Cambridge Schools Examinations Board

13 Draw a polygon ABCDEF making AB = 32 mm, BC = 38 mm, CD = 50 mm, DE = 34 mm, EF = 28 mm, FA = 28 mm, AC = 56 mm, AD = 68 mm and AE = 50 mm. Construct a further polygon similar to ABCDEF but having an area larger in the ratio of 4:3.

Cambridge Local Examinations

14 Construct, full size, the figure illustrated in Figure 5.71 and, by radial projection, superimpose about the same centre a similar figure whose area is three times as great as that shown in Figure 5.71.

Oxford Local Examinations

15 Figure 5.72 shows a section through a length of moulding.

Draw an enlarged section so that the 118 mm dimension becomes 172 mm.
Oxford Local Examinations

16 Figure 5.73 shows a shaped plate, of which DE is a quarter of an ellipse.
Draw:

(a) the given view, full size;
(b) an enlarged view of the plate so that AB becomes 150 mm and AC
    100 mm. The distances parallel to AB and AC are to be enlarged
    proportionately to the increase in length of AB and AC, respectively.
Oxford Local Examinations

17 Construct a regular hexagon having a distance between opposite sides
of 100 mm. Reduce this hexagon to a square of equal area. Measure and
state the length of side of this square.

Joint Matriculation Board

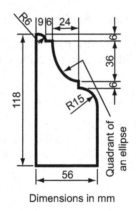

Dimensions in mm

Figure 5.72

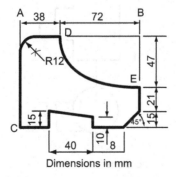

Dimensions in mm

Figure 5.73

18 A water main is supplied by two pipes of 75 mm and 100 mm diameter. It is required to replace the two pipes with one pipe which is large enough to carry the same volume of water.

Part 1. Draw the two pipes and then, using a geometrical construction, draw the third pipe.
Part 2. Draw a pipe equal in area to the sum of the three pipes.
Southern Regional Examinations Board

19 Three squares have side lengths of 25, 37.5 and 50 mm, respectively. Construct, without resorting to calculations, a single square equal in area to the three squares, and measure and state the length of its side.

Cambridge Local Examinations

Intersection of Regular Solids

20 Figure 5.74 shows the plan and incomplete elevation of two cylinders. Draw the two views, showing any hidden lines.

North Western Secondary School Examinations Board

21 The plan and incomplete elevation of two pipes are shown in Figure 5.75. Copy the two views, full size, and complete the elevation showing hidden detail.

Middlesex Regional Examining Board

22 Figure 5.76 shows an incomplete elevation of the junction of a cylinder and an equilateral triangular prism. The axes of both lie in the same VP. The prism rests with one of its side faces in the HP.

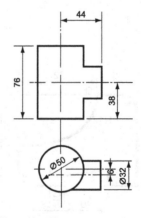

Dimensions in mm

Figure 5.74

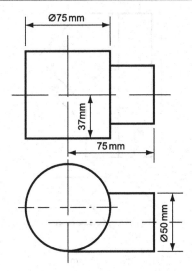

Figure 5.75

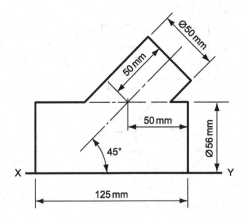

Figure 5.76

Draw, and complete, the given elevation and project a plan. Do *not* show hidden detail.

Joint Matriculation Board

23 Figure 5.77 shows incomplete drawings of the plan and elevation of a junction between a square section pipe and a cylindrical pipe.

Draw (a) the complete plan and elevation and (b) the development of the whole surface of *either* the square pipe *or* the cylindrical one.

Southern Universities' Joint Board

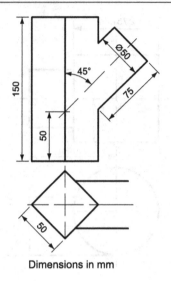

Dimensions in mm

Figure 5.77

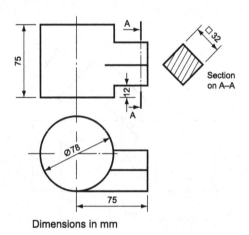

Dimensions in mm

Figure 5.78

24 Figure 5.78 gives the plan and incomplete elevation of a junction between a cylinder and a square prism. Copy the two views and complete the elevation showing all hidden detail.

Oxford and Cambridge Schools Examination Board

25 Figure 5.79 consists of a plan and incomplete elevation of a square prism intersecting a cone.

(a) Draw the given plan.
(b) Draw a complete elevation showing the curve of intersection.

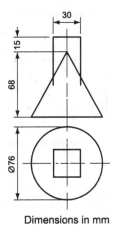

Dimensions in mm

Figure 5.79

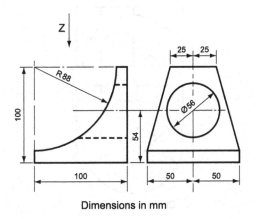

Dimensions in mm

Figure 5.80

(c) Develop the surface of the cone below the curve of intersection.
Southern Universities' Joint Board

26 Two elevations of a wheel stop are shown in Figure 5.80. Draw, full size, (a) the given elevations, (b) a plan looking in the direction of arrow Z.

Associated Examining Board

27 Figure 5.81 shows an incomplete plan view of the junction of a cylinder to a right square pyramid. The axis of the cylinder is 32 mm above the base of the pyramid that stands on the HP. Perpendicular height of the pyramid = 100 mm.

Draw and complete the given plan and project an elevation looking in the direction of arrow A.

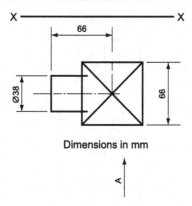

Dimensions in mm

Figure 5.81

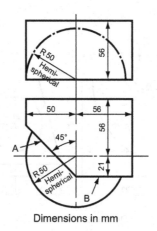

Dimensions in mm

Figure 5.82

Draw, also, the development of the pyramid portion showing the hole required to receive the cylinder. Show all hidden detail.

Joint Matriculation Board

28　The height of a right circular cone is 88 mm and the base diameter is 94 mm. The cone is pierced by a square hole of side 32 mm. The axis of the hole intersects the axis of the cone 32 mm above the base and is parallel to the base.

Draw an elevation of the cone looking in a direction at right angles to the vertical faces of the hole.

Oxford and Cambridge Schools Examination Board

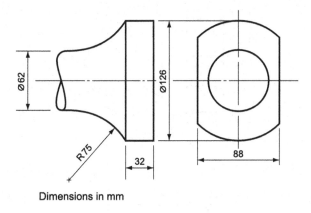

Dimensions in mm

Figure 5.83

29 The plan and incomplete elevation of a solid are shown in Figure 5.82. Reproduce the given views and complete the elevation by including the lines of intersection produced by the vertical faces A and B. Hidden edges are to be shown.

   Cambridge Local Examinations

30 Two views (one incomplete) of a connecting rod end are shown in Figure 5.83. The original diameters were 126 mm and 62 mm and the transition between these followed a circular path of radius 75 mm. Two flat parallel faces were then milled as shown in the EE. Draw the given views, complete the left-hand elevation and beneath this project a plan. Scale: full size.

   Oxford Local Examinations

Developments

31 Two views and an isometric view of a cement mixer cover are given in Figure 5.84. Using a scale of $\frac{1}{12}$, draw the two given views and add an EE. Then, using the same scale, draw the development of the sheet steel needed to make this cover.

   Southern Regional Examination Board

32 Figure 5.85 shows the plan and elevation of a tin-plate dish. Draw the given views and construct a development of the dish showing each side joined to a square base. The plan of the base should be part of the development.

   Middlesex Regional Examining Board

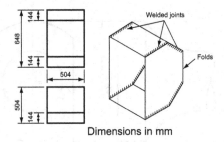

Dimensions in mm

Figure 5.84

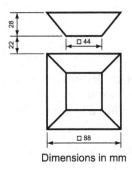

Dimensions in mm

Figure 5.85

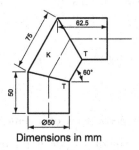

Dimensions in mm

Figure 5.86

33   Figure 5.86 shows three pipes, each of 50 mm diameter and of negligible thickness, with their axes in the same plane and forming a bend through 90°. Draw:

(a) the given view;
(b) the development of pipe K, using TT as the joint line.
Associated Examining Board

34 An FE of the body of a small metal jug is given in Figure 5.87. Draw, full size, the following:

(a) the front view of the body as shown;
(b) the side view of the body looking in the direction of arrow S;
(c) the development of the body with the joint along AB.
South-East Regional Examinations Board

35 Two views of a solid are given in Figure 5.88. Determine the development of the curved surface of the solid.

Oxford and Cambridge Schools Examination Board

36 The plan and elevation of a thin metal sheet are shown in first angle projection in Figure 5.89. A bar D of diameter 38 mm is placed on the

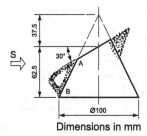

Dimensions in mm

Figure 5.87

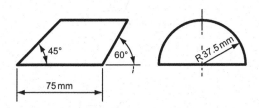

Figure 5.88

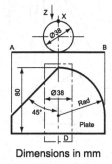

Dimensions in mm

Figure 5.89

plate which is then tightly wrapped round the bar so that edges A and B of the plate meet along a line at X. Draw a plan view of the wrapped plate when looking in the direction of arrow Z, assuming that the bar has been removed.

Cambridge Local Examinations

37 Figure 5.90 shows the elevation and partly finished plan of a truncated regular pentagonal pyramid in first angle projection.

(a) Complete the plan view;
(b) develop the surface area of the sloping sides.
Cambridge Local Examinations

38 Figure 5.91 shows two views of an oblique regular hexagonal pyramid. Draw, full size:

(a) the given views;

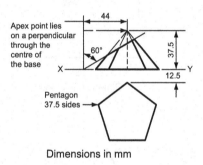

Dimensions in mm

Figure 5.90

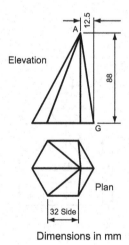

Dimensions in mm

Figure 5.91

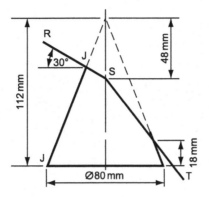

Figure 5.92

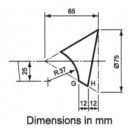

Dimensions in mm

Figure 5.93

(b) the development of the sloping faces only, taking 'AG' as the joint line. Show the development in one piece.

Associated Examining Board

39 Draw the development of the curved side of the frustum of the cone, shown in Figure 5.92, below the cutting plane RST. Take JJ as the joint line for the development.

Associated Examining Board

40 Make an accurate development of the sheet metal adaptor piece which is part of the surface of a right circular cone as shown in Figure 5.93. The seam is at the position marked GH.

Cambridge Local Examinations

Chapter 6

Views

6.1 BASIC CONVENTIONS

So far, the previous chapters have outlined the principles of orthographic projection and how to represent an object in multi-view drawings. Remember, depending on where the object is positioned with respect to the horizontal plane (HP) and the vertical plane (VP), either the first angle projection method or the third angle projection method is used. If the object is above the HP and in front of the VP, then first angle projection is used. Accordingly, if the object is below the HP and behind the VP, then third angle projection is applied (compare section 4.5).

Obviously, orthographic projection can result in a variety of representations of the object, depending on the relative spatial position between object and projection plane. However, some views are more important and more convenient than others.

Imagine the object, to be shown as a multi-view drawing, is located in a glass box. The glass box itself consists of six mutually perpendicular planes which could serve as projection planes (Figure 6.1). This leads to six principal views. Usually, an object can be oriented in the glass box in such a way that one or more major surfaces are parallel to the glass box, which results in projections on the six principal planes, showing the most important features of interest.

6.1.1 The Six Principal Views

As presented in Chapter 4, to obtain the views according to the third angle projection method, we need to hinge the planes (glass box side walls) and unfold them so that all planes are in line. Since in third angle projection the object is below the horizontal plane (HP) and behind the front vertical plane (FVP), we need to fix the FVP and unfold all other planes. The result is shown in Figure 6.2. This is the standard arrangement of views in third angle projection.

The front view usually is the central view and shows most of the part's features, ideally in true size. All other resulting views are automatically derived from the aforementioned process of unfolding. That means, the top view is located above the front view, the bottom view is below the front view, the right-side view is located on the right-hand side of the front view and, similarly, the

DOI: 10.1201/9781003001386-6

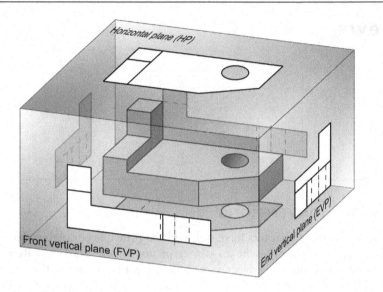

Figure 6.1 Glass box model for an object shown in third angle projection.

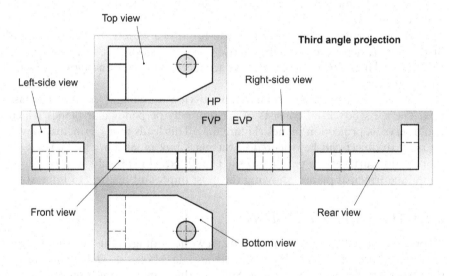

Figure 6.2 Unfolded glass box and the resulting six principal views in third angle projection.

left-side view can always be found on the left-hand side of the front view. If a rear view is necessary, it can be placed either to the left of the left-side view or to the right of the right-side view.

As can be seen from Figure 6.2, the features are not shown in true size in every view. For example, the true shape of the inclined face is not visible in any of the given principal views. However, its true height (measured perpendicular to the HP) can be taken from the front or right-side view. To see the true width of that face, an auxiliary view would be necessary.

Per definition, a maximum of two of the three principal dimensions (width, height, depth) can be shown in one view at the same time. The remaining dimension(s) can be taken from adjacent views.

If the unfolding is properly done, all views will be automatically aligned with each other. This is also mandatory to ensure that any misinterpretation is avoided. In manual drafting, it is possible to complement any incomplete views by projecting entities from one view into another.

Although the allowable number of views on an engineering drawing is not limited, many drawings come up with three views: the front view, the top view, and either the right-side view or left-side view. These views are sometimes called the 'regular views'.

In Chapter 4, the alternative orthographic projection method in first angle is introduced. To receive the six principal views in first angle projection, the same glass box from Figure 6.1 is unfolded. Since in first angle projection the object is located above the HP and in front of the FVP, the unfolding of the planes leads to a different arrangement of views (Figure 6.3).

The front view is still the central element, but now the top view is located below the front view, while the bottom view is above the front view. Accordingly, the right-side view is on the left-hand side of the front view and the left-side view on the right-hand side of the front view.

To avoid any confusion, the projection method shall be indicated, usually by a small symbol, located in or close to the title block and which represents a truncated cone in two views (Figure 6.4).

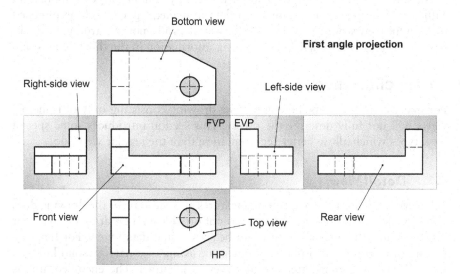

Figure 6.3 Unfolded glass box and the resulting six principal views in first angle projection.

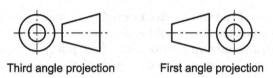

Third angle projection   First angle projection

Figure 6.4 Graphical symbol for third and first angle projection method according to ISO 5456–2.

6.1.2 Selection of Views

A frequently occurring question is *How many of which views do I need to fully describe an object?* There is no single answer to this question since the number of required views and the selection of views highly depends on the part's geometry and which dimensions to show. However, some general guidelines should be followed whenever drafting.

Any engineering drawing shall be lean and legible. For that reason, the drawing should only contain views which are absolutely necessary to fully describe the object without ambiguity, or which contribute to an understanding of the part's function.

As a rule, the front view is selected in such a way that it is most descriptive.

Based on the selected front view, further projected views have to be chosen to clearly represent the individual features of the part to be shown.

Take care of inclined and oblique faces, and select appropriate views. If necessary add auxiliary views to show features in true size.

In case of manufacturing drawings, sometimes an object needs to be drawn in the orientation of its main manufacturing position. As an example, for a part manufactured on a lathe, the centre line would be drawn horizontal.

For simple parts, a single view supplemented by annotations or symbols can suffice. This applies, for example, to flat sheet metal parts such as punched plates. One view would fully describe the outline, and a note regarding the thickness of the part would deliver the missing information in the third dimension.

6.2 SPECIAL VIEWS

For many complex parts, an engineering drawing consisting only of principal views will not fully describe the object. This section introduces some special view types which allow better representation than the regular views do.

6.2.1 Detail View

Whenever large parts comprise small features and the scale of the drawing does not allow legible representation or does not provide enough drawing space for dimensioning, the unclear feature can be shown in a detail view. For that, the feature of interest is enclosed by a circle consisting of a narrow solid line and marked with a capital letter which serves as identifier. The encircled area is drawn on an enlarged scale so that all necessary dimensions can be placed. The

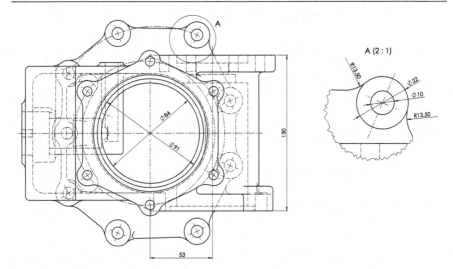

Figure 6.5 Detail view for a flange hole of a gearbox housing.

detail view can be placed anywhere on the drawing sheet as long it is clearly marked by the identification letter and the scale used for that view.

Figure 6.5 shows an example. The main view of the presented gearbox housing is shown at a scale of 1:1, but the scale is too small to fully dimension a hole of the integral flange. On top of the view, one hole is encircled and marked with the letter A. Now the detail can be drawn at a scale of 2:1 so that the dimensions can be placed. The letter height for indicating the view and the scale shall be one size larger than the standard letter height. Since in mechanical engineering the letter height usually is 3.5 mm, the letter height for the view identifier and the scale is 5 mm.

6.2.2 Partial Views

Views do not need to be shown in full if only a certain detail is of interest. Partial views can be used to clearly describe a feature of interest without showing the full view. This can save sketching time but can also lead to increased legibility because unnecessary parts of the view are not shown.

A partial view can be one of the six principal views or any other view type. A narrow freehand line limits the partial view.

Figure 6.6 illustrates how a partial view of an engine block can replace a space-consuming full view.

6.2.3 Broken Views

Whenever objects need to be drawn to show large differences in their principal dimensions, e.g. length to width, then this would usually require either using a

Full view

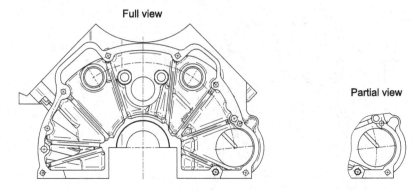

Partial view

Figure 6.6 An engine block shown as a full view and a partial view.

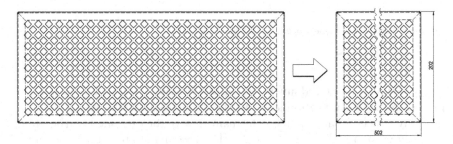

Figure 6.7 Broken view of a punched frame.

small scale or increasing the drawing sheet size. A typical example is a long pipe or beam. Showing the part in full length could lead to a large portion of empty drawing space. To avoid this, the part can be shown shortened. That means that the part is drawn at the desired scale but a middle portion is left out.

Figure 6.7 depicts an example of a punched frame. Owing to its length, the part would probably not fit on A4 or A3 paper and still show all the details at a satisfying scale. On the right-hand side the punched frame is shown as a broken view, i.e. the middle portion is left out and the view is interrupted by narrow freehand lines. Note: the overall length is dimensioned as it would not have been broken, which means the dimension shall show the true length.

6.2.4 Auxiliary Views

Some objects show features whose faces are not parallel to any of the glass box side walls and therefore cannot be shown in true size in any of the six principal views. The solution is to introduce an auxiliary view in such a way that the auxiliary plane is parallel to the feature's major face. The development of auxiliary elevations is explained in depth in Chapter 4. Therefore, at this point it suffices

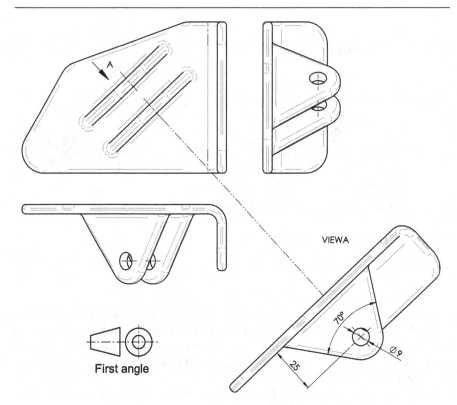

Figure 6.8 Angle bracket shown in front view, left-side view, top view and auxiliary view.

to point out that auxiliary views can be helpful to clearly describe an object. Figure 6.8 depicts an example shown in first angle projection.

It is worth mentioning that complex objects require successive auxiliary views, which means, for example, that a secondary auxiliary view is projected from a primary view.

Since auxiliary views are sometimes difficult to interpret due to their non-standard orientation, hidden lines should be omitted to increase legibility. Of course, hidden edges can be shown if they are absolutely necessary and contribute to the understanding of the object's shape.

If there is a risk that the arrangement of views could lead to misinterpretation, a viewing-plane line can be used, as shown in Figure 6.8. Alternatively, a direction arrow and a capital letter as view identifier may be used. The arrow indicates the direction of sight for the projected auxiliary view.

6.2.5 Removed Views

A removed view is not a new view type, but it designates a view which is positioned at another place on the drawing sheet. As a consequence, this view is no longer in direct projection, i.e. aligned, with other views.

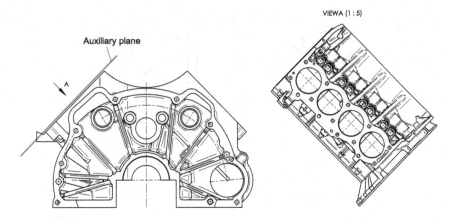

Figure 6.9 Auxiliary view as a removed view.

One reason to remove views can be a lack of drawing space.

In principle, any view of any view type can be removed. Since the view is not aligned anymore, it is difficult to figure out from which direction of projection the view is derived from. For that reason, removed views need to be marked by a letter for identification and the direction of projection is to be indicated in any parent view.

Figure 6.9 shows an engine block given in its front view. A necessary auxiliary view to see the holes for the pistons would lead to a significant increase of required drawing space since it would be projected perpendicular to the auxiliary plane. To save space, the auxiliary view is shown removed and therefore not aligned to its parent view anymore. The view identifier and the arrow clearly indicate the relationship between the front view and the auxiliary view.

6.3 SECTIONAL VIEWS

Sections have already been discussed at some length in Chapter 4, where their main application was in finding the true shape across a body. When sections are used in engineering drawing, although the true shape is still found, the section is really used to show what is inside the object.

A drawing must be absolutely clear when it leaves the drawing board or the CAD program. The person or persons using the drawing to make the object must have all the information that they need presented clearly and concisely so that they are not confused – even over the smallest point.

Suppose that you had to draw the assembly of the three-speed gearbox on the rear hub of a pedal cycle. You probably know nothing about the interior of that hub. The reason that you know nothing about it is that you cannot see inside it. If you are to produce a drawing that can be read and understood by anybody, you can draw as many views of the outside as you wish, but your drawing will still tell nothing about the gear train inside. What is really needed is a view of the inside of the hub, and this is precisely what a section allows you to show.

6.3.1 Conventions for Sections

Figure 6.10 shows two sections projected from a simple bracket. For clarity, the drawing additionally shows an isometric view of the bracket including tangent lines. You will note that the given sections are both projected from the FE. Sections can be projected from any elevation – you are not limited to the FE only. Thus, you can project a sectioned FE from either the plan or the EE. A sectioned EE is projected from the FE and a sectioned plan is projected from the FE. It is not usual to project a sectioned EE from the plan or vice versa.

The lines A–A and B–B are called the sectional cutting planes, and this is a good description because, in fact, you are pretending to cut the bracket right through along these lines. Both the sectioned EEs are what you would see if you had physically cut the brackets along A–A and B–B, removed the material *behind* the cutting planes (i.e. the side away from the arrowheads) and projected a normal EE with the material removed. To avoid any misinterpretation, and to show the section quite clearly, wherever the cutting plane has cut through material the drawing is hatched.

The standard hatching for sectioning is at 45°, although it will be seen later that in exceptional circumstances this rule may be broken. As an example, hatching lines should not be drawn parallel or perpendicular to an outline. You should also note that the cutting planes pass through holes in the bracket and they are not hatched. Hatching should only be done when the cutting plane passes through a solid material.

Hatching lines shall be uniformly spaced and neither too close nor too distant. The distance between hatching lines can be adapted to the overall size of the

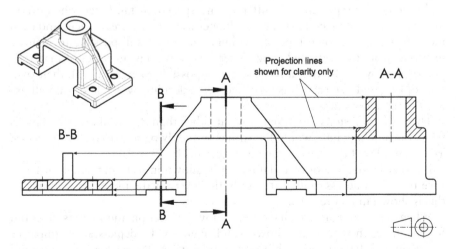

Figure 6.10 Two sections projected from a simple bracket.

drawing. However, make sure that the hatching of a part is always the same in all views, with respect to hatching angle and distance between the hatching lines.

The lines A–A and B–B are of a particular nature. According to ISO 128–3, they are dashed dotted wide lines. The letters A–A and B–B are not a random choice either. Sectional cutting planes are usually given letters from the beginning of the alphabet, although you will sometimes see other letters used.

Sectioning is a process which should be used only to simplify or clarify a drawing. You should certainly not put a section on every drawing that you do. There are some engineering details that, if sectioned, lose their identity or create a wrong impression, and these items are never shown sectioned. A list of these items are shown as follows.

Nuts and bolts	Ball bearings and ball races
Studs	Roller bearings and roller races
Screws	Keys
Shafts	Pins
Webs	Gear teeth

Webs are not shown sectioned because section lines across a web give an impression of solidarity and lead to confusion. Compare Figure 6.10 where the web is excluded from the hatching in section view B–B although the web is cut by the cutting plane.

The question of clarity arises again when considering an assembly, i.e. more than one part. If any of the parts are in the list provided earlier, they are not hatched, but a finished product may be composed of several different parts made with several different materials. In the days when productivity was not quite so vital, the draftsperson was someone who turned out drawings that were almost works of art. Since there was no printing as there is today, only one drawing was made. Each different material was coloured when sectioned and each colour represented a different and specific metal. Later, when drawings were duplicated, colours were no longer used to any great extent and each metal was given its own type of shading, and it was still possible to identify materials from the sectioned views. There are now so many types of materials and their alloys in use that it has become impossible to give them all their own type of line, and you can please yourself when deciding which line you will use for a particular section.

There are occasions when hatching at other than 45° is allowable. This is when the hatching lines would be parallel or nearly parallel to one of the sides. Two examples of these are shown in Figure 6.11.

If a very large piece of material has to be shown in section, then, in order to save time, it is necessary to hatch only the edges of the piece. An example of this is shown in Figure 6.12.

When very thin materials have to be shown in section and there is no room for hatching, then they are shown solid. Figure 6.13 depicts an example of two mounted flanges with a thin gasket in between. The gasket is shown solid black.

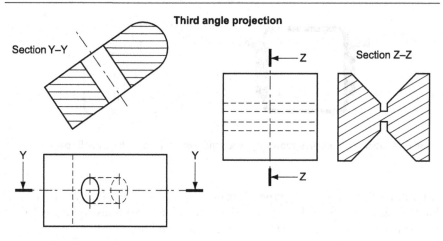

Figure 6.11 Hatching lines other than 45°.

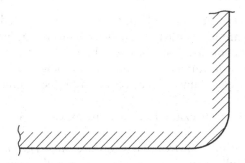

Figure 6.12 Hatching lines for large bodies.

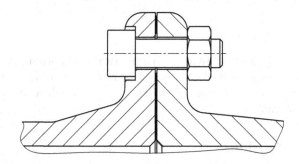

Figure 6.13 Sectioning of thin materials.

The most common occurrence for thin parts, of course, is when drawing sheet metal. If two or more parts are shown adjacent, a small space should be left between them. As an example, in Figure 6.14 three structural members are shown in a sectional view. The isometric view with tangent lines is shown for clarity only.

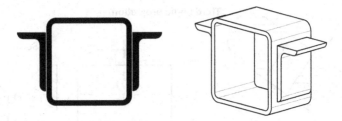

Figure 6.14 Structural profiles shown in a sectional view, separated by a small space.

Omit hidden lines in any type of section view unless absolutely necessary. If hidden lines need to be shown, note that a sectioned area is always bounded by visible edges or a centre line but never by a hidden line.

6.3.2 Full Sections

Whenever an object is cut fully in half, the view is designated as a full section view. Full sections are used either for single parts to reveal interior features or for how the components are assembled and how they interact with each other.

If there is no risk of ambiguity, a full section view can be shown without indicating a cutting plane line. The direction of projection must be obvious to the reader.

Figure 6.15 depicts an example of a gearbox casing shown in first angle projection.

On the right-hand side, the front view of the casing is given. The direction of projection for the full section view is obvious since the view is aligned to the front view. The cutting plane for the section view must be in line with the vertical centre line indicating the centre of the front view.

6.3.3 Half Sections

If the object that you are drawing is symmetrical and nothing is to be gained by showing it all in section, then it is necessary to show only as much section as the drawing requires. This usually means drawing a half section. Figure 6.16 depicts an example.

6.3.4 Offset Sections

Sometimes interior features of a part which are not aligned to each other require a section view. This would make several section views necessary. The number of views can be kept to a minimum by using offset section views instead of multiple full sections.

The example illustrated in Figure 6.17 shows a casted cover in front view. The cover comprises several holes in a total of eight different sizes. Most of

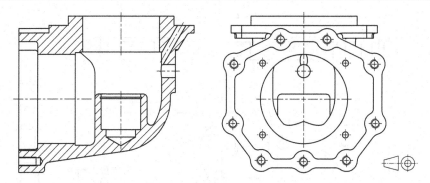

Figure 6.15 Full section view of a gearbox casing without indication of cutting plane.

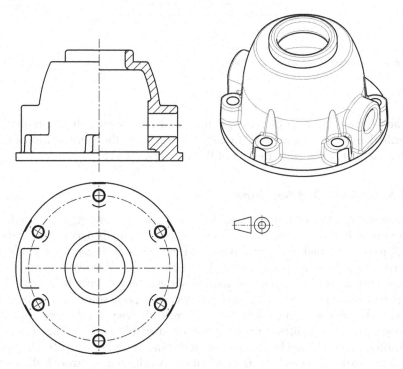

Figure 6.16 Casing shown in top view and as a half section.

them are not aligned and would require adding multiple section views. Six of these holes are exemplarily shown in two offset section views.

The path of the cut shall be represented by a narrow dashed dotted line and the thick cutting line is only drawn at the end of each cutting plane and whenever the direction of the cutting plane changes.

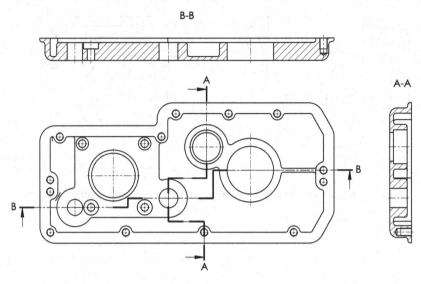

Figure 6.17 Casted cover with two offset sections.

Each offset cut requires a designation, and the viewing direction shall be indicated by reference arrows. In each section view the identification of the referenced cut is placed at the top of the view (Figure 6.17).

6.3.5 Broken-Out Sections

Remembering that sections are used only to clarify a drawing, it is quite likely that you will come across a case where only a very small part of the drawing needs to be sectioned to clarify a point. In this case, a broken-out section is permitted. The sectioned area is bounded by a continuous narrow freehand line.

In Figure 6.18, a transmission shaft is shown. Since shafts belong to those parts which are never shown completely sectioned, one solution to expose the interior shapes is to add broken-out sections to the view. In this example, they clearly represent the different tapped holes and enable proper dimensioning.

Another example of a broken-out section is shown in Figure 6.19. The example depicts a housing of a centrifugal pump. A full section through the centre line of the outlet would expose the correct interior shape of the outlet, but it would also provide a wrong impression of the exterior shape. By showing the housing in a regular view and adding a broken-out section, the interior as well as the exterior shape can be seen without the risk of misinterpretation.

6.3.6 Aligned Sections

Whenever a detail needs to be shown but is not located along any regular cutting plane, an aligned section may be applied. For that, the cutting plane is bent

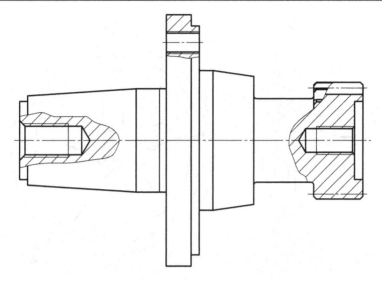

Figure 6.18 Shaft with three local sections.

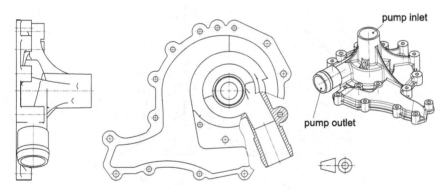

Figure 6.19 Broken-out section in a view from a housing of a centrifugal pump.

so that it passes through the detail of interest. The section view finally shows all sectioned areas aligned to one single plane, i.e. all features are revolved into the same plane.

Figure 6.20 provides an example of a hand wheel with five spokes. Any straight cutting plane would lead to a sectional view that shows only one spoke at a time. The cutting plane is bent in such a way that it passes through two spokes.

Note, aligned sections should be deliberately used since they can lead to an impression of distortion. In Figure 6.21 a fixture is shown with three equally distributed lugs. The aligned section view gives the false impression of the

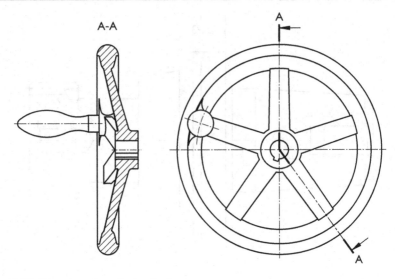

Figure 6.20 Aligned section of a hand wheel.

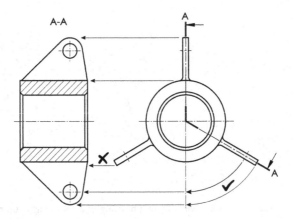

Figure 6.21 Aligned section of a fixture with lugs.

overall height. Therefore, the aligned section should be used to dimension specific features, whereas principal dimensions shall be shown in other views.

6.3.7 Profile Sections

It is often necessary to show a small section showing the true shape across an object. There are two ways of doing this, and they are both shown in Figure 6.22. The revolved section is obtained by revolving the section in its

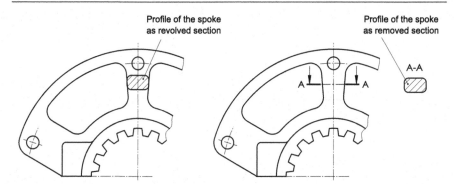

Figure 6.22 Profile of a spoke shown as revolved section and as removed section.

position by 90° and breaking the outline to accommodate the section. The removed section is self-explanatory and should be used instead of the revolved section if there is room on the drawing, as it is much neater.

Chapter 7

Dimensioning

7.1 INTRODUCTION

In contrast to pictorial representations in user manuals, spare part catalogues and assembly instructions, an engineering drawing is not an illustration. An engineering drawing, or technical drawing, differs from a technical illustration in that it is strongly regularised by standards and represents a detailed specification of the size and shape of a single part or of an assembly. The subjects of interest are the dimensions and associated tolerances.

When an engineering drawing is made, dimensioning is of vital importance. All the dimensions necessary to manufacture a part must be on the drawing and they must be presented so that they can be easily read, easily found and not open to misinterpretation. A neat drawing can be spoilt by bad dimensioning.

The principles for presentation of dimensions that apply to any kind of 2D engineering drawing is internationally standardised in ISO 129–1. National standards may differ from the given conventions.

Figure 7.1 explains the structure of dimensions in engineering drawings. In general, dimensions consist of a dimension line, extension lines and a dimension value. This applies to linear dimensions and angular dimensions as well. An extension line is sometimes also referred to as a *projection line* and represents an extension of a feature outline or centre line. Extension lines indicate the starting and end point of a dimension. They are drawn as narrow continuous lines and extend from the outline to approximately 2 mm past the dimension line. While according to international standards the extension line shall touch the outline or feature to which it refers, Anglo-American standards require a visible gap between them. The dimension line shall be a narrow line according to ISO 128–2 (compare Chapter 2) and terminate with filled arrowheads as long as the letter height (usually 3.5 mm), and these arrowheads must touch the extension lines from the inside, as can be seen in Figure 7.2 (a).

Alternative terminators are closed but unfilled arrowheads, open arrowheads (either 30° or 90° opening angle), oblique strokes and dots. If possible, only one type of arrowhead should be used on a drawing. If there is space for the dimension text but not the arrowheads, the numerical value is placed between the extension lines, whereas the arrowheads are placed outside the extension lines (Figure 7.2 (b)). When neither the numerical value nor the arrowheads can

DOI: 10.1201/9781003001386-7

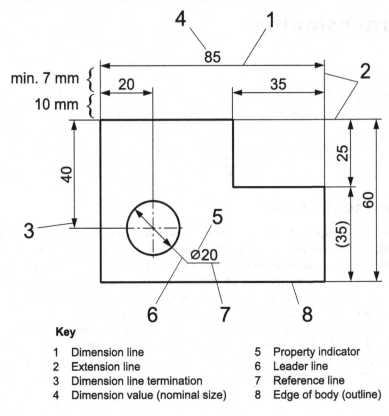

Figure 7.1 Structure of dimensions.

Key

1	Dimension line	5	Property indicator
2	Extension line	6	Leader line
3	Dimension line termination	7	Reference line
4	Dimension value (nominal size)	8	Edge of body (outline)

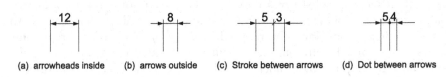

(a) arrowheads inside   (b) arrows outside   (c) Stroke between arrows   (d) Dot between arrows

Figure 7.2 Use of line terminations.

be placed between the extension lines, both can be placed outside. Strokes and dots can be used in case of limited space between extension lines (Figure 7.2 (c) and (d)).

The distance between an object's outline and the first dimension shall be 10 mm. Spacing between two parallel dimensions is minimum 7 mm. It is essential to find the balance between compact representation of an object and legibility of the drawing. Therefore, sometimes the mentioned distances can be increased for the sake of clarity and legibility. On the other hand, when drawing space is limited but legibility still given, the distances can be kept to a minimum.

7.2 DIMENSIONAL VALUES

Lettering used for dimensions shall be in accordance with ISO 3098, where character height is ten times the width of a narrow line. If appropriate, dimension values are preceded by a 'property indicator', which is a symbol that describes a feature's shape and its associated dimension type. Examples are the diameter symbol ⌀ for circular or cylindrical features, or the letter 'R' to indicate a radius dimension. Further property indicator symbols are presented in Table 7.1.

Where used, the property indicator shall precede the dimensional value without space. Figure 7.3 shows some examples on how to use property indicators

Table 7.1 Property indicators for dimensions

Indicator	Denotation	Application
□	Square	Square feature with equidistant sides described by one side dimension.
R	Radius	Circular or cylindrical feature described by its radius.
SR	Spherical radius	Spherical feature described by its radius.
⌒	Arc length	Dimension indicating the curvilinear length of a curve segment.
⌀	Diameter	Circular or cylindrical feature described by its diameter.
S⌀	Spherical diameter	Spherical feature described by its diameter.
⊔	Counterbore	Cylindrical flat-bottomed hole described by its diameter and depth.
∨	Countersink	Conical hole described by its diameter and angle.
↧	Depth	Depth of an internal feature, e.g. hole.
▷	Conical taper	Conical taper described by its taper ratio.
◺	Slope	Flat taper defined by its slope ratio.

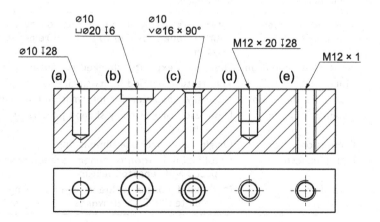

Figure 7.3 Property indicators used for e.g. dimensioning of machined holes.

and depicts different types of machined holes with their respective callouts. The hole callouts have the following meaning:

(a) Blind hole of diameter 10 mm and depth equal 28 mm. Note, the depth does not include the bit and therefore refers to the length of the cylindrical feature only.

(b) Hole of diameter 10 mm with a counterbore of diameter 20 mm and a depth of 6 mm.

(c) Hole of diameter 10 mm with a countersink. The diameter at the surface equals 16 mm and the included angle is 90°.

(d) Tapped hole with a metric thread of size 12 mm. Thread length equals 20 mm, whereas the depth of the core hole is 28 mm.

(e) Tapped hole with a metric fine thread of size 12 mm and a pitch of 1 mm.

The presented property symbols are also used in dimensioning according to ASME standards. Therefore, the hole callouts presented in Figure 7.3 can be internationally understood.

Each dimension shall be given on the drawing only once. If it is necessary to show the dimension a second time for clarity, the second dimension shall be given in parentheses which indicates an auxiliary dimension. This kind of dimension, sometimes also referred to as *reference dimension*, is for information purpose only and will not be considered for production. Therefore, auxiliary dimensions are never toleranced.

A technical drawing may require further special dimensions. These dimensions have a complementary indication, providing additional details to fully describe a part. Following ISO 129–1, Table 7.2 gives an overview of the indicators used in technical drawings to further specify special dimensions.

Table 7.2 Complementary indicators for dimensions

Indicator	Denotation	Application
□	Theoretical exact dimension	A feature's size, location, orientation or profile dimensioned relative to a datum reference system.
⬯	Inspection dimension	Dimension explicitly checked for quality control.
t =	Thickness	Thin feature defined by its thickness.
( )	Reference dimension	Auxiliary dimension for information purpose only.
[ ]	Rough dimension	Dimension of a primary shaped part before machining.
⊙	Dimension origin	Origin for coordinate dimensions.
×	Multiplication	Simplified indication of dimensions for equally spaced and repeated features.
_	Out of scale	Modified dimension value in paper drawing. Prohibited in CAD drawings.

Engineering drawings can be shown in either millimetres (abbreviated as *mm*) or inches (abbreviated as *in* or sometimes also as "). Most countries outside the United States use the metric system of measure. However, due to globalisation and internationally operating companies, it might be possible that a dual unit system is used. In that case, dimensions are given in both millimetres and inches.

When using the metric system, dimensions are usually given in millimetres, irrespective of the size of the dimension. However, in other disciplines such as civil engineering the metre is used, sometimes also the centimetre. If the majority of linear dimensions on a drawing is of the same unit, it is not necessary to include the abbreviation for the unit used, i.e. cm or mm. Instead, the unit system is indicated on an engineering drawing, inside or close to the title block. For angular dimensions, the unit shall be given with each individual dimension.

Depending on the applied standard, the decimal separator is shown as a dot, i.e. 15.26, or as a comma, i.e. 15,26 or 0,003. While countries that are using the British imperial unit system or US customary system prefer the dot, ISO 129–1 suggests a comma as decimal marker. Values less than unity are prefixed by a nought.

As a rule, dimensional values should not be defined arbitrarily. Instead, values from series of preferred numbers should be used whenever possible. The usage of preferred numbers leads to advantages in product development and manufacturing such as reduction of part variety and assurance of part exchangeability due to standardisation. Many standard machine elements, plate thicknesses, tolerances, or radii follow a series of preferred numbers, as for example defined in the international standard ISO 3. The basic series R 5, R 10, R 20 and R 40 are rounded members of geometric series, i.e. following a geometric progression. The members of a basic series are obtained by multiplying the previous member by a rounded factor, the *gradation*. These multipliers q_n for the basic series according ISO 3 are calculated as follows:

$$R\,5: \quad q_5 = \sqrt[5]{10} \approx 1.6$$

$$R\,10: q_{10} = \sqrt[10]{10} \approx 1.25$$

$$R\,20: q_{20} = \sqrt[20]{10} \approx 1.12$$

$$R\,40: q_{40} = \sqrt[40]{10} \approx 1.06$$

Note that each series of higher order root includes the values of the preceding series and some intermediate values for refinement. Series R 5 is preferred over R 10, R 10 over R 20 and R 20 over R 40. In technical drawings, preferred numbers are used for various applications. One example is the gradation of radii values. To avoid arbitrary selection, radii should be selected from the basic R 10 series:

0.2 0.4 0.6 1.0 1.6 2.5 4 6 10 16 20 25 32 40 50 ...

Similarly, thread sizes are standardised and follow series of preferred numbers. As an example, the basic sizes for coarse metric ISO threads follow the series R 5 and R 10:

R 5 :   M 2.5   M 4   M 6   M 10   M 16   M 24   M 36   ....

R 10 :   M 3   M 5   M 8   M 12   M 20   M 30   ...

7.3 SYSTEMS OF DIMENSION POSITIONING

Dimensional values can be either placed parallel, or *aligned*, to their dimension line so that they can be read from the bottom or the right of the drawing sheet (ISO standard) or simply *unidirectional* so that they can always be read from the bottom (ASME standard). While unidirectional dimensions are easier to read, aligned dimensioning allows conservation of drawing space. Technical drawings should never show a mixture of the two different dimension positioning systems. Use one of the two systems consistently throughout the drawing. Usually the required reading direction(s) is given by company internal rules. Figure 7.4 exemplarily shows a dimensioned part with different dimension positioning systems. Within this book, aligned dimensioning is exclusively applied.

In aligned dimensioning – dimensions not oriented towards any of the two main reading directions, the dimensional value is still aligned to the dimension line in such a way that it can be read from the bottom or from the right-hand side (see Figure 7.5).

Whenever possible, dimensions should be placed in the most descriptive view for the feature of interest. At the same time, dimensions should not be placed within the feature's contour.

Never dimension hidden features unless absolutely necessary and not prone to misinterpretation. As a rule, there is another view where the feature of interest can be clearly shown so that dimensioning hidden edges can be avoided.

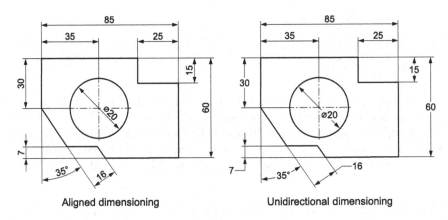

Aligned dimensioning   Unidirectional dimensioning

Figure 7.4 Comparison of aligned and unidirectional dimensioning.

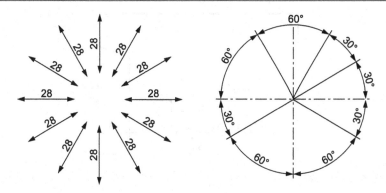

Figure 7.5 Reading directions.

7.4 ARRANGEMENT OF DIMENSIONS

Dimensions can be placed in a variety of combinations. When creating an engineering drawing, it is crucial to arrange dimensions in such a way that functional aspects as well as requirements regarding inspection and manufacturing are taken into account. In the following, the different methods of arranging dimensions are presented. As a rule, there is no standard method to apply. Instead, the recommended arrangement depends on the specific part and the intended use.

Chain Dimensioning

In this method of arrangement, dimension lines of different features are aligned in such a way that they form a chain as shown in Figure 7.6. With chain dimensioning, the location of a feature is based on the previous feature, therefore its positional accuracy depends on the exact positioning of the predecessor. Series of chained dimensions should be avoided whenever possible, because tolerances can accumulate and finally lead to malfunctioning of the part or assembly. This problem of accumulating tolerances will be discussed in detail in the next chapter. Note, when chain dimensioning is required, for example for manufacturing reasons, adding an overall dimension, as shown in the example of Figure 7.6, can lead to over-dimensioning. For that reason, either the overall length or one of the feature dimensions shall be shown as an auxiliary dimension.

Parallel Dimensioning

By applying this method, it is possible to overcome the drawbacks of chain dimensioning. The individual dimension lines are placed parallel to one another, i.e. stacked or aligned one adjacent to the other and referring to a common extension line as a common datum. Figure 7.7 exemplarily shows a drawing of a plate with parallel linear dimensions and some concentric angular dimensions. Note, the common reference for all vertical linear dimensions is the

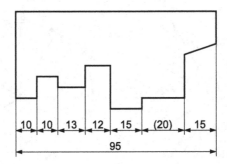

Figure 7.6 Chain dimensioning.

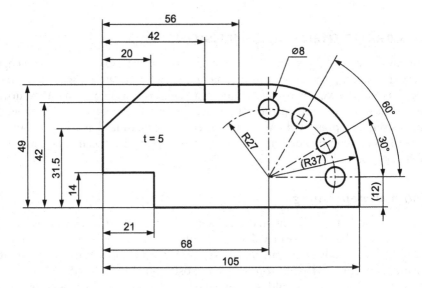

Figure 7.7 Parallel dimensioning applied to a plate.

bottom edge, whereas the common reference for all horizontal linear dimensions is the leftmost edge. The horizontal centre line of the pitch circle is used as a datum for the angular dimensions.

Compared to other methods of dimension arrangements, parallel dimensioning provides a better positional accuracy since possible accumulation of tolerances is avoided. Instead, the allowable deviations are given for each individual dimension, which leads to exact positioning. As a rule, the overall dimensions are given but not as auxiliary dimensions.

When using the method of parallel dimensioning, it might be possible that a drawing needs more than just one common datum for the horizontal or vertical dimensions. This can have multiple reasons which are usually related to the function of the part or related to inspection or manufacturing. As an example,

Step 1
Chucking the workpiece

Step 2
Turning right shoulders of the shaft

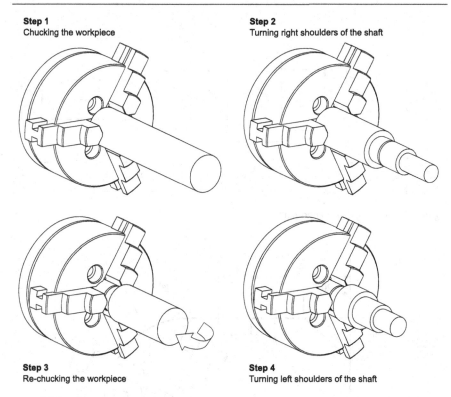

Step 3
Re-chucking the workpiece

Step 4
Turning left shoulders of the shaft

Figure 7.8 Turned shaft in a three-jaw chuck.

parts which are machined on a lathe usually require minimum two references for the horizontal dimensions. Figure 7.8 depicts a three-jaw chuck which is used on a lathe to fix the part to be machined. First, all shoulders accessible from one side are turned while the part is fixed in the chuck. After re-chucking, the shoulders on the other side of the part are turned.

For that reason, all shoulders of one side are referring to the same datum, the end face of the shaft. The shoulders on the other side of the shaft refer to the opposite end face. Figure 7.9 shows the resulting technical drawing for the given part.

There is one drawback when applying parallel dimensioning. A lot of drawing space is required compared to other methods. In addition, dimensioning features from a common reference is not always suitable, especially from a manufacturing or an inspection point of view. For that reason, a combination of parallel dimensioning and chain dimensioning is commonly used.

Combined Dimensioning

With this arrangement, requirements derived from the intended function of the part, from manufacturing or from inspection, can be considered in a better way.

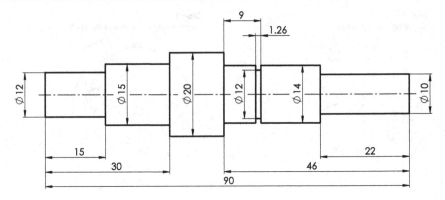

Figure 7.9 Drawing of a turned part.

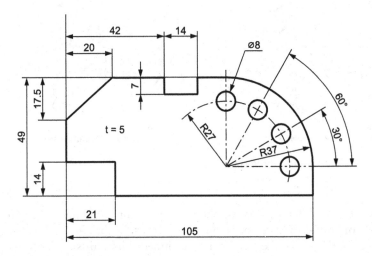

Figure 7.10 Combined dimensioning.

Figure 7.10 shows an example of combined dimensioning, applied to the plate introduced in Figure 7.7.

The decision of whether dimensions are arranged as a chain or refer to a common datum should never be based on aiming for a reduction of drawing space. Instead, the arrangement should solely depend on aspects considering the three main criteria: function, manufacturing and inspection.

Superimposed Running Dimensioning

This kind of arrangement eliminates the drawback of parallel dimensioning and reduces the space needed for placing dimensions. In contrast to parallel dimensioning, all dimension lines of the same direction are superimposed. Therefore, as a rule, only one single dimension line is used for each direction. Each origin

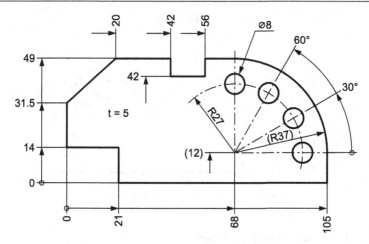

Figure 7.11 Superimposed running dimensioning applied to a plate.

is indicated by a small circle and the dimensional value zero. Further dimensions are added by placing the dimensional value near the terminator, aligned with the respective extension line. Figure 7.11 depicts an example of superimposed running dimensioning, applied to the plate shown in Figure 7.7.

To avoid confusion by extension lines completely crossing the view, dimensions can also be shown on a shortened dimension line with one arrowhead directing to the feature of interest. As an example, see dimensions 20, 42 and 56 shown on top of the view in Figure 7.11.

In the presented example, unidirectional running dimensions are applied since the main origin is located in the lower left corner of the part. In case of bidirectional running dimensions, two sets of dimension lines aligned in opposite direction need to be used. While one set is pointing in the positive direction, the set of dimensions pointing in the opposite direction will indicate negative dimension values and therefore must provide a minus sign (see Figure 7.12).

The arrangement of superimposed dimensions is mainly used for numerical controlled (NC) machined parts, e.g. turning, milling and drilling.

Coordinate Dimensioning

Similar to superimposed running dimensioning, features refer to a common origin, but with this method only one overall origin is used. Size dimensions and location dimensions are defined by Cartesian coordinates, starting at the common origin and defined by linear dimensions in orthogonal directions. The individual coordinate values are entered near the coordinate points in the given view or entered separately in a table (see Figure 7.13).

If coordinate values are given in a table, coordinate points shall be designated by a numerical or alphanumerical reference letter for clear identification.

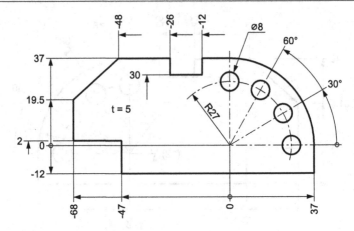

Figure 7.12 Bidirectional running dimensioning.

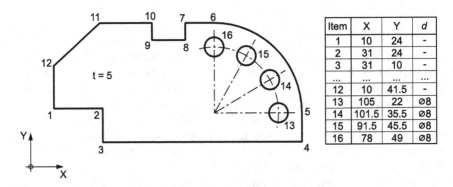

Item	X	Y	d
1	10	24	-
2	31	24	-
3	31	10	-
...	...	...	...
12	10	41.5	-
13	105	22	⌀8
14	101.5	35.5	⌀8
15	91.5	45.5	⌀8
16	78	49	⌀8

Figure 7.13 Cartesian coordinate dimensioning.

The reference letter can be placed adjacent to the coordinate point within the object's outline or be placed outside the outline by using a leader line.

If features such as holes need to be dimensioned by providing an angle from a reference direction and a distance from a reference point, polar coordinates can be used. In polar coordinate dimensioning, an origin is defined and the angle 0° starts on the right-hand side of the origin, located on the polar axis. Positive angles are measured counter-clockwise. Figure 7.14 depicts an example of a drawing where polar coordinates are combined with unidirectional superimposed dimensioning. Alternative to showing the holes with visible edges, the holes' position can be represented by crossing lines.

Since no dimension lines or extension lines are used, coordinate dimensioning reduces the required drawing space. As a rule, coordinate dimensioning is applied for NC machined features and parts. This method is well suited for these applications because the dimensions are individually toleranced and accumulation cannot occur.

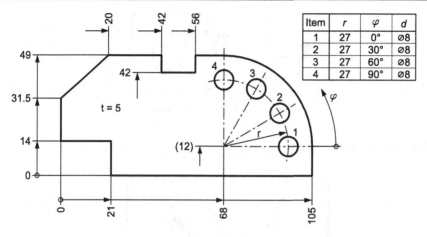

Item	r	φ	d
1	27	0°	⌀8
2	27	30°	⌀8
3	27	60°	⌀8
4	27	90°	⌀8

Figure 7.14 Polar coordinate dimensioning combined with superimposed running dimensions.

7.5 PRESENTATION OF SPECIAL DIMENSIONS

In engineering drawing, some geometric entities frequently appear in a drawing view. This section presents typical features and how to dimension them.

Radii

Whenever a part is machined, casted or injection-moulded, fillets and rounds occur. In the derived drawing views, these features usually appear as arcs. A circular arc is always dimensioned in the view where its true shape is given. The dimensional value is preceded by the property indicator R and the dimension line points to the centre point of the arc. Only one arrowhead is used as line termination and is located at the arc, either inside or outside the feature's outline. Inside the outline is preferred, if possible. The centre point of an arc shall be marked by two crossing narrow lines if the exact location is necessary to indicate for functional reasons. If the centre point of an arc falls outside the drawing, the dimension line can be shown shortened with a broken dimension line. If arcs of the same size are repeated in a drawing view, it is sufficient to dimension just one arc if unambiguity is given. Alternatively, the dimension lines can be shown combined. If all arcs shown on a drawing are of the same size, a note close to the title block can be given, such as 'All fillets and rounds R5 unless otherwise specified'. Figure 7.15 shows possible drawing entries for arcs.

Diameters

Circular shapes are to be dimensioned with a preceding diameter symbol ⌀. A circle can be dimensioned by drawing a dimension line between two extension lines, projected from two diametrically opposite points on the circle. Alternatively, if sufficient space is given, the circle can be dimensioned internally,

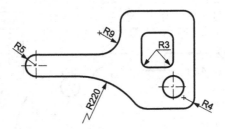

Figure 7.15 Dimensioning radii.

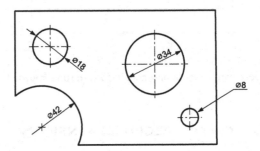

Figure 7.16 Dimensioning diameters.

i.e. the dimension line crosses the centre point of the circle. When the circle is too small to place the dimensional value inside the circle, the value can be placed outside. Possible variations of dimensioning diameters are shown in Figure 7.16.

Squares and Widths Across Flats

Sometimes parts contain a square feature. If the side lengths including their tolerance are identical, dimensioning can be simplified by dimensioning just one side and using the property indicator for a square □. Note, the dimension shall only be placed in a view where the feature can be clearly identified as a square or in an orthogonal view. Whenever a given flat face might be misinterpreted as a cylindrical face, it shall be marked with two crossing diagonals, drawn as narrow continuous lines (see Figure 7.17).

To avoid adding a supplementary view just for dimensioning two opposing flat faces, the width across flats can be indicated with a leader line. The dimensional value for the distance is preceded by the upper case letters WAF (width across flats). Figure 7.18 depicts an example of how to dimension a width across flats. Note, the pictorial view is given for clarity only.

Chamfers

A chamfer is a bevelled edge, usually added to two adjacent faces to avoid sharp edges. They are used, for example, to ensure the function of two mating

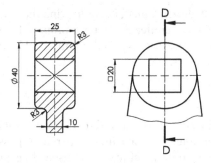

Figure 7.17 Dimensioning square features.

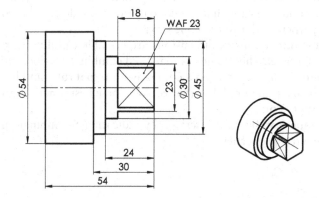

Figure 7.18 Dimensioning a width across flats.

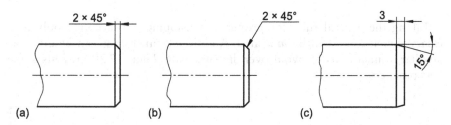

(a)   (b)   (c)

Figure 7.19 Dimensioning a chamfer.

parts, to avoid injuries and to prevent damage to the part. As a rule, a chamfer is created at an angle of 45°. However, other angles might be necessary for specific applications. When dimensioning a chamfer, the length of the offset and the angle shall be provided. As long as the angle is 45°, the presentation can be simplified according Figure 7.19 (a). When drawing space is limited or the value for the offset is very small, the dimension can be given with a leader line as shown in Figure 7.19 (b). Whenever the angle is unequal 45°, the chamfer shall be conventionally dimensioned, i.e. indicating the angle and separately the offset or alternatively the chamfer diameter (compare Figure 7.19 (c)).

The examples shown are for external chamfers. However, the same rules apply for internal chamfers.

Round and Flat Tapers

Frequently, conical surfaces, also referred to as *tapers*, can be found at mechanical parts. As an example, with a round taper it is possible to establish an efficient shaft-hub joint to fix machine elements such as pulleys, sprockets and couplings on the end of a shaft. Round tapers require special attention when it comes to dimensioning. For simplicity, the presented dimension technique refers to truncated cones but can also be applied to any type of cone. Truncated cone in this context means that the cone is a right-angle circular cone and cut by a plane perpendicular to the cone's axis.

Different combinations of dimensions are possible to fully describe the size and shape of a taper. There is no standardised combination since ultimately the choice of dimensions depends on a company's internal rules and how the taper is going to be manufactured. For that reason, some best practices are presented with no claim to completeness.

A round taper can be fully described by indicating a combination of the following geometric properties:

- Taper ratio C,
- cone angle α,
- larger cone diameter D,
- smaller cone diameter d, and/or
- length L.

Taking the general rules for proper dimensioning into account, only necessary dimensions shall be indicated. Additional dimensions can be given as auxiliary dimensions to avoid over-dimensioning. Figure 7.20 presents some best practices.

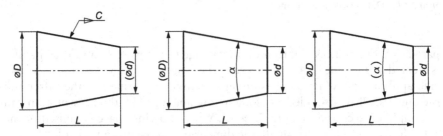

Figure 7.20 Best practices on how to dimension a round taper.

When indicating the taper ratio C, this represents the ratio of the difference in the two diameters D, d and the length L. The ratio is calculated by the following formula:

$$C = \frac{D-d}{L} = 2 \cdot \tan\left(\frac{\alpha}{2}\right)$$

Note, when using the taper ratio, the property indicator ⊳ is entered before the taper ratio value and placed on a reference line which points to the taper with the help of a leader line. The orientation of the taper symbol shall match the direction of the taper on the workpiece.

Tapers are not limited to round tapers. Instead, pyramidal tapers can also occur. The presented methods of dimensioning and the definition of the taper ratio can also be applied to pyramidal tapers if the diameters are substituted by dimensional values describing the rectangular cross-sections of the pyramid. Figure 7.21 depicts examples of tapers and the indication of the taper ratio.

Another type of machine element which can be used to connect a shaft to other machine elements is a tapered key. In contrast to round tapers, this kind of flat taper has a one-sided inclined surface. Flat tapers are dimensioned in a very similar way to round tapers, except that the property indicator ⊱ is used. Apart from the different preceding symbol, the same rules apply as those for round tapers, i.e. the orientation of the symbol shall match the orientation of the inclined surface of the workpiece and the taper ratio indicates the slope (Figure 7.22).

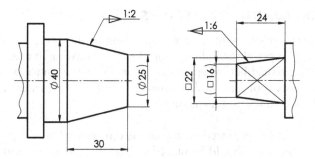

Figure 7.21 Orientation of the taper symbol.

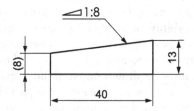

Figure 7.22 Dimensioning a flat taper.

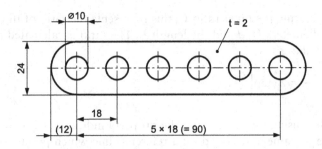

Figure 7.23 Equally spaced features in linear direction.

Equally Spaced and Repeated Features

Sometimes parts consist of a regular pattern of features, i.e. they are equally spaced in a given direction. In such case, dimensioning can be simplified by fully dimensioning the first feature of the pattern, dimensioning the space between the first and the second feature and finally dimensioning the distance between the first and the last feature of the pattern. The latter dimension consists of a value indicating the number of patterned elements, the distance between two instances of the pattern, and the sum of the linear or angular spacing given as an auxiliary dimension. Figure 7.23 and Figure 7.24 provide examples of equally spaced features and how to dimension them.

7.6 DIMENSIONING GUIDELINES

There are many rules about how to dimension a drawing properly, but it is unlikely that two people will dimension the same drawing in exactly the same way. However, remember when dimensioning that you must be particularly neat and concise, thorough and consistent. The following rules (Figure 7.25) must be adhered to when dimensioning:

1   The dimension is placed on top of the dimension line.
2   The dimensions should be placed so that they are read from the bottom of the paper or from the right-hand side of the paper.
3   Dimension lines should be drawn *outside* the outline, whenever possible, and should be kept well clear of the outline.
4   Overall dimensions should be placed outside the intermediate dimensions.
5   To improve readability of a drawing, dimensions for internal features and external features shall, wherever possible, be arranged and indicated in separate groups of dimensions. Figure 7.25 shows an example.
6   If dimensions cannot be shown due to limited space between the extension lines, dimension lines may be extended past the extension lines and the arrowheads positioned outside of them, in reverse.

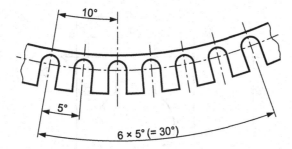

Figure 7.24 Equally spaced features in angular direction.

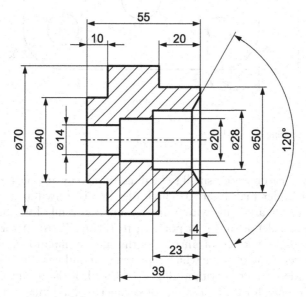

Figure 7.25 Illustrating rules 1-6.

7 Intersections between dimension lines and any other line shall be avoided. If an intersection is unavoidable, the dimension line is shown unbroken while the other line will be shown interrupted.

8 Centre lines and outlines must *never* be used as dimension lines. They may be used as extension lines.

9 Diameters may be dimensioned in one of two ways. Either dimension directly across the circle (*not* on a centre line) or project the diameter to outside the outline (Figure 7.26).

10 When dimensioning a radius, if possible you must show the centre of the radius. The actual dimension for the radius may be shown on either side of the outline but of course should be kept outside if possible.

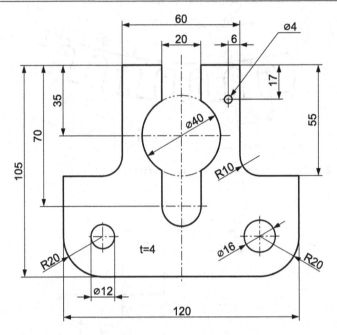

Figure 7.26 Illustrating rules 9-11.

11 When a feature is too small to be dimensioned by any of the prior methods, a leader may be used. The leader line is drawn as a narrow continuous line and ends with a line termination symbol depending on the addressed detail that it is pointing to (Figure 7.26). If the leader line ends on the surface of an object, the line termination is a filled dot. When referring to a line which represents an outline or an edge of a part, the leader line ends with a filled arrowhead. If the leader line refers to another line, for example a line of symmetry or a dimension line, no line termination is used. Long leader lines should be avoided even if it means inserting another dimension. The leader line should always meet another line at an acute angle.

12 Dimensions shall be given in a view which shows the feature of interest most clearly. As an example, usually holes can be identified and dimensioned without ambiguity where the axial view on it is given.

13 Dimensions should *not* be repeated on a drawing. A common mistake is over-dimensioning. If indicating the same dimension is necessary, show it as an auxiliary dimension, i.e. in parentheses.

14 Unless unavoidable, do *not* dimension a hidden detail. It is usually possible to dimension the same detail on another view.

15 When dimensioning angles, draw the dimension lines with a compass; the point of the compass should be on the point of the angle. The arrowheads may be drawn on either side of the dimension lines, and the dimension may be inserted between the dimension lines or outside them.

Whatever the angle, the dimension must be placed so that it can be read from either the bottom of the paper or from the right-hand side.

16 If many parallel dimensions are given, it avoids confusion if the dimensions are staggered so that they are all easier to read. A portion of the dimension line can even be omitted if unambiguity is given such as in the case of diameters (Figure 7.27). Dimension lines from diameter dimensions may also be shown not in full if only one-half of a symmetric part is given or if an object is shown as a half-section view (Figure 7.28).

17 If the drawing is to scale, the dimensions put on the drawing are the actual dimensions of the component and not the size of the line on your drawing.

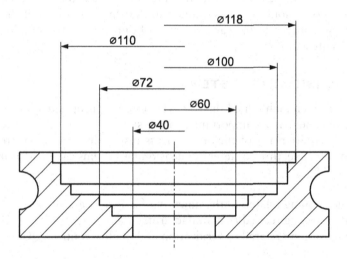

Figure 7.27 Staggered diameter dimensions.

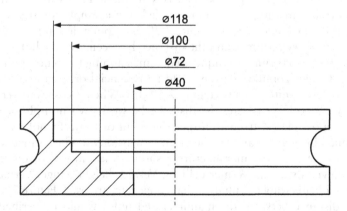

Figure 7.28 Diameter dimension lines shown not in full.

The aforementioned rules do not cover all aspects of dimensioning (there is a new set on toleranced dimensions alone), but they should cover the essential ones. Dimensioning properly is a matter of applying common sense to the rules because no two different drawings can ever raise exactly the same problems. Each drawing that you do needs to be studied very carefully before you begin to dimension it.

Examination questions often ask for only five or six 'important' dimensions to be inserted on the finished drawing. The overall dimensions – length, breadth and width – are obviously important but the remaining two or three are not so obvious. The component or assembled components need to be studied in order to ascertain the function of the object. If, for instance, the drawing is of a bearing, then the size of the bearing is vitally important because something has to fit into that bearing. If the drawing is of a machine vice, then the size of the vice jaws should be dimensioned so that the limitations of the vice are immediately apparent. These are the types of dimensions that should make up the total required.

7.7 DIMENSIONING SYSTEMS

So far it has been highlighted that the selection of appropriate dimensions and their arrangement is an important process step. It is crucial to consider the needs and requirements of the departments responsible for downstream processes. In principle, three different dimensioning systems can be identified (also see Figure 7.29):

- Dimensioning related to the function of a part or assembly
- Dimensioning related to manufacturing
- Dimensioning related to verification and quality control (QC)

When a part is designed, usually the intended function of a part is considered. However, it might be necessary to dimension the part from a manufacturing point of view so that it can be easily manufactured without calculating or transforming missing dimensions needed, for example for the machining process. In the example given in Figure 7.29 (a), from a design point of view it might be necessary to dimension the distance between the two holes. Depending on the manufacturing technique used, dimensioning the centre points of the two holes from a common datum could be beneficial or even necessary from a manufacturing point of view (Figure 7.29 (b)). When it comes to verification or quality control, measuring the distance between two non-physical centre points is impossible. So from an inspection point of view, the holes need to be dimensioned in such a way that the given dimensions can be verified, compare Figure 7.29 (c). In many manufacturing shops, conventional or digital measurement devices such as Vernier calliper, Vernier height gauge, Vernier depth gauge, outside micrometre, plunger dial gauge, etc. are used. Figure 7.30 shows how the distance between the manufactured holes would be verified with a Vernier calliper according to Figure 7.29 (c).

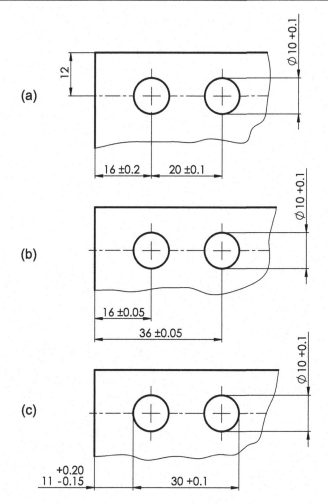

Figure 7.29 Systems of dimensioning.

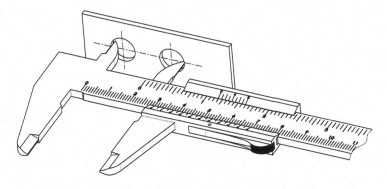

Figure 7.30 Verifying a dimension with a Vernier calliper.

Chapter 8

Tolerancing

8.1 DIMENSIONAL TOLERANCES

8.1.1 Introduction

With the advent of manufacturing machines during the Industrial Revolution came demands for precisely manufactured parts. Up to that time, companies were already striving for interchangeability of parts. After some disillusioning attempts to introduce techniques to guarantee dimensional accuracy, the first tolerances for manufactured parts were established in the mid-19th century. This was the beginning of interchangeable manufacturing, i.e. parts could be produced in such a way that the accuracy was sufficient to permit assembly and the intended function without further reworking or accepting high scrap rates.

Nowadays, parts are frequently manufactured at different manufacturing facilities and at different times. However, if they are assembled they need to fit each other at all times. With the introduction of standardisation, best practices on how to produce certain parts, such as frequently occurring machine elements, were well documented. The introduction of tolerances allowed manufacturers to be able to work according these standards.

Any manufacturing process will lead to smaller or larger deviations from the ideal dimension given on the technical drawing. Therefore, it must be specified when a part quality from a dimensional point of view is still acceptable or not. This is solved by applying dimensional tolerances. Hence, a tolerance describes the permissible level of deviation.

To understand the principle of tolerancing, some definitions of size designation according to ISO 286–1 are provided as follows (also see Figure 8.1):

- Nominal size N: ideal (theoretical) size as specified in the technical drawing
- Actual size I: measured size after manufacturing
- Upper limit of size ULS: maximum permissible size
- Lower limit of size LLS: minimum permissible size
- Deviation: actual size minus the nominal size

DOI: 10.1201/9781003001386-8

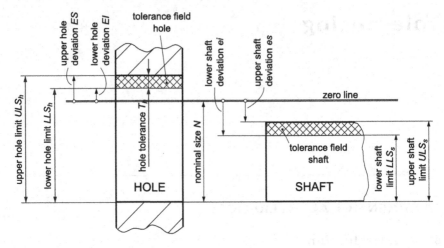

Figure 8.1 Definitions for size designation.

- Upper limit deviation: upper limit of size minus nominal size
  - designated *ES* for internal features
  - designated *es* for external features
- Lower limit deviation: lower limit of size minus nominal size
  - designated *EI* for internal features
  - designated *ei* for external features
- Tolerance *T*: difference between upper and lower limit of size or upper and lower limit deviation

The upper and lower limit deviations are designated by abbreviations derived from their French translation (*ES*/*es* = écart supérieur; *EI*/*ei* = écart inférieur).

Note, by convention the term *shaft* is used to designate all external features whose size needs to be measured from outside (e.g. shafts, blocks, plates). Likewise, the term *hole* is used to designate all internal features whose size needs to be measured from inside (e.g. cylindrical holes, grooves, rectangular cut-out).

The *zero line* represents the nominal size (= zero deviation), therefore any deviation is measured from this theoretical line. The tolerance field, also referred to as *tolerance interval*, is defined by the tolerance limits, i.e. the upper and lower limit of size of the feature. Since the allowable upper and lower deviations can be either negative, positive or zero, the tolerance interval does not necessarily contain the zero line. Upper and lower deviation both can be positive, upper deviation can be positive while the lower deviation is negative, and both deviations can be negative (Figure 8.2). A tolerance interval is characterised by its thickness and the position relative to the zero line.

The presented definitions of size designation can be internationally understood. However, in the American standard ASME Y14.5M two additional

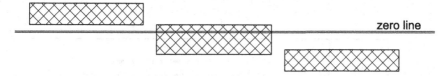

zero line

Figure 8.2 Possible positions of a tolerance interval.

terms are introduced. The condition of a part when it consists of the greatest amount of material is designated as *maximum material condition* (MMC). This means that for an external feature of size MMC is given by the upper limit while for an internal feature it is given by its lower limit. Similarly, the condition of a part when it consists of the least possible amount of material is named *least material condition* (LMC). For external features it equals the lower limit, for internal features it equals the upper limit.

8.1.2 General Tolerances

In fact, all dimensions on a drawing need a specification of their allowable limits. To avoid having to give individual deviations for each dimension, general tolerances are applied. This eases the drafting process, and generally defined tolerances can take commonly used manufacturing machines' precision into account. General tolerances can be given on a drawing as a note or within the title block.

A standardised classification of general tolerances is given in ISO 2768–1. According to this standard, general tolerances are classified in four degrees:

- f → fine
- m → medium
- c → coarse
- v → very coarse

For each *tolerance class*, deviations are provided in equal bilateral form, i.e. in symmetrical form, within different given size ranges. In mechanical engineering, usually the tolerance class 'medium' is used. Therefore, the given deviations consider accuracy typical for the shop floor. The class 'fine' is applied in precision engineering such as designing microscopy components and watch movement units. Tolerance class 'coarse' is used for foundry technology and, for example, agricultural engineering. Very coarse deviations are not used in practice anymore since modern manufacturing machines are capable of achieving a much higher precision.

Determining the allowable deviations for a given nominal size requires looking up the desired degree and the respective size range that the nominal size fits in. As an example, Figure 8.3 shows how to determine 'medium' deviations for a feature's size of 25 mm.

②

Tolerance class	Linear dimensions						
	Limit deviations in mm for nominal size ranges						
	0.5 to 3	over 3 to 6	over 6 to 30	over 30 to 120	over 120 to 400	over 400 to 1000	...
f (fine)	± 0.05	± 0.05	± 0.1	± 0.15	± 0.2	± 0.3	...
m (medium)	± 0.1	± 0.1	± 0.2	± 0.3	± 0.5	± 0.8	...
c (coarse)	± 0.2	± 0.3	± 0.5	± 0.8	± 1.2	± 2	...
v (very coarse)	-	± 0.5	± 1	± 1.5	± 2.5	± 4	...

①

③

Tolerance class	Radii and chamfers			Angular dimensions			
	Limit deviations in mm for nominal size ranges			Limit deviations in degrees and minutes for nominal size ranges			
	0.5 to 3	over 3 to 6	over 6	to 10	over 10 to 50	over 50 to 120	...
f (fine)	± 0.2	± 0.5	± 1	± 1°	± 0° 30'	± 0° 20'	...
m (medium)							
c (coarse)	± 0.4	± 1	± 2	± 1° 30'	± 1°	± 0° 30'	...
v (very coarse)				± 3°	± 2°	± 1°	...

Figure 8.3 Looking up general tolerances.

8.1.3 Specific Tolerances

Whenever a general tolerance does not provide a desired accuracy, a specific tolerance is used. While general tolerances are not shown on a technical drawing for each individual dimension, specific tolerances need to be indicated with the respective dimension. Specific tolerances always override any given general tolerances.

The deviations for a specific tolerance can be shown in different ways, as depicted in Figure 8.4 for linear dimensions and in Figure 8.5 for angular dimensions. Deviations are always given succeeding the dimensional value. When deviations are to be presented in unequal bilateral form or unilateral form, the letter height is one size smaller than the dimension height. Since usually the line group 0.5 is used, dimensions are shown with a line thickness of 0.35 mm and a letter height 3.5 mm (10 × line thickness). Deviations are therefore shown with a letter height of 2.5 mm. When the upper and lower deviation are equal, the tolerance is symmetrical and can be given in equal bilateral form. In this case the deviation is shown in the same letter height as the dimensional value.

The selection of appropriate tolerances during the design process is a very important step. Special attention is needed since part tolerances usually have a big impact on the selection of the manufacturing technique, on the assembla-bility of two mating parts and on the performance of the assembly. Overall, tolerances notably influence product costs. As can be seen from Figure 8.6, dimensional tolerances have an exponential effect on manufacturing cost. Taking the tolerances which can be achieved with casting or forging as a reference, the costs remain approximately constant as long as we have tolerances down

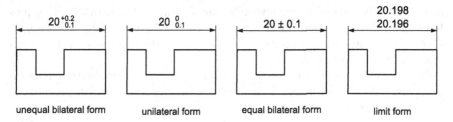

unequal bilateral form   unilateral form   equal bilateral form   limit form

Figure 8.4 Possible linear tolerancing.

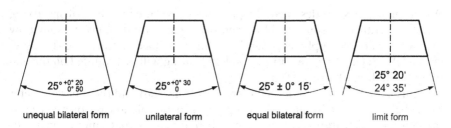

unequal bilateral form   unilateral form   equal bilateral form   limit form

Figure 8.5 Possible angular tolerancing.

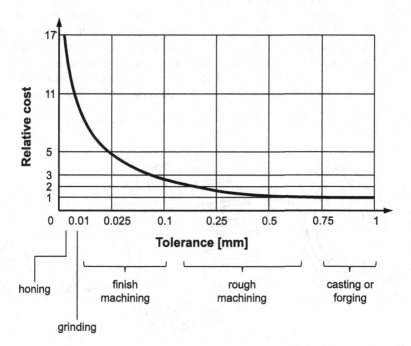

Figure 8.6 Qualitative representation of the dependence of manufacturing cost on tolerances.

to ~0.5 mm. When requesting finer tolerances, other manufacturing techniques have to be chosen, which results in higher costs. Manufacturing technologies such as grinding or even honing and lapping can easily lead to costs ten times higher than the reference.

Owing to the significance of selected tolerances and their direct impact on costs, the following principle shall always be applied: *As coarse as possible, as fine as necessary.*

8.2 FITS

8.2.1 Introduction

When single parts are designed, the previously presented method of specifying specific tolerances is suitable. However, when parts are designed to somehow mate to each other, applying general tolerances or specific tolerances becomes critical. As an example, an antifriction ball bearing will be mounted on the end of a pinion drive shaft (see Figure 8.7).

Let us assume that the nominal diameter of the shaft and the inner diameter of the ball bearing is 20 mm. When applying general tolerances (tolerance class medium), the deviations are ±0.2 mm. Therefore, an actual inner diameter of the ball bearing of e.g. 19.85 mm and an actual shaft diameter of e.g. 20.2 mm would be within the permissible size range. Although each actual size would satisfy the condition, it would not be possible to assemble the two parts since the actual diameters result in an interference.

If, instead of general tolerances, specific tolerances are applied, the deviations for the hole of the ball bearing indicate its minimum and maximum permissible diameter. It also determines the cost to manufacture the inner cylindrical face with the given accuracy. However, the tolerance does not consider the shaft at all. The deviations specified for the shaft indicate the minimum and maximum permissible diameter of the respective shaft shoulder. They also determine the

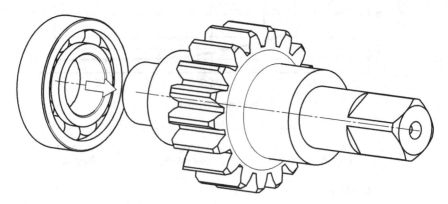

Figure 8.7 Two parts to be mated.

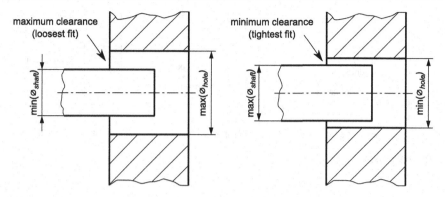

Figure 8.8 Maximum and minimum clearance between two mating parts.

manufacturing cost, but ultimately the shaft tolerance does not consider the hole of the ball bearing at all.

Obviously, considering individual tolerances is not suitable to enable two parts to be successfully mated. Instead, it is necessary to consider the relationship between the tolerances of the two parts. This kind of relationship is expressed by a so-called *fit*.

A fit describes the degree of tightness or looseness between two mating parts. When the specific tolerances for two mating parts are given, only the boundaries of the permissible size ranges are considered. Depending on the defined deviations, the pair of tolerances can result in a more or less loose or tight fit (see Figure 8.8). The loosest fit is given when the largest possible hole diameter minus the smallest possible shaft diameter results in a maximum clearance. The loosest fit corresponds to the *tolerance* of the two mating parts. On the other hand, the tightest fit is given when the smallest possible hole diameter minus the largest possible shaft diameter results in a minimum clearance or even in an interference. The tightest fit is also referred to as *allowance*.

8.2.2 Types of Fits

By specifying permissible deviations for two mating parts, i.e. defining two tolerance intervals with well-defined limits, three different types of fit can result (Figure 8.9).

Clearance Fit

Taking the lower and upper limit into account, the internal mating part will always be smaller than or equal to the external mating part. A clearance fit is characterised by a positive allowance, i.e. there is always a small clearance between the smallest possible hole and the largest possible shaft. Depending on the amount of clearance, this fit type can be further subdivided into running

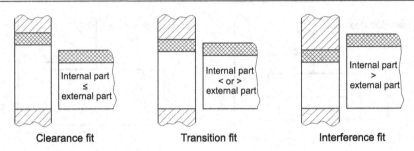

Figure 8.9 Types of fit.

fits, sliding fits and locational clearance fits. This type of fit is used when rotating or sliding between the mating parts is necessary, e.g. movable clutches, sealing rings, sliders, pillar guides, and so on.

Transition Fit

If the two tolerance intervals overlap each other so that the internal mating part could be smaller or larger than the external mating part, then this is a transition fit. Depending on the actual sizes of the two mating parts, it can result in a very small clearance or in a small interference. Transition fits are used when accurate location is required. Examples are machine elements on a shaft such as gears, pulleys and bearings.

Interference Fit

If the internal mating part will always be larger than the external mating part, then this is an interference fit. Since usually we have to apply a certain force to assemble two mating parts in an interference fit, this is also called *force fit* or *shrink fit*. As an example, whenever forces need to be transmitted from one component to the other, interference fits are used. Examples are bushings in a wheel hub, clutches on a shaft end and rocker arms.

Within each fit category it is possible to define size limits so that the resulting fit becomes looser or tighter. The selection of a fit during the design stage needs special attention since accurate fits result in higher manufacturing cost. Tighter fits usually require more effort for assembly. As an example, in case of interference fits special tools are needed.

It seems that a more detailed classification is necessary if we want to further specify the degree of clearance or interference. With ISO 286–1, a code system for tolerances on linear sizes is introduced.

8.2.3 The ISO Code System

Since it is theoretically possible to specify arbitrary deviations for two mating parts, the code system presented by the International Organization for

Standardization allows a reduction of possible combinations of tolerance intervals. In addition, interchangeability can be achieved by standardising values for tolerances and deviations. Instead of indicating lower and upper limits for two mating parts on a drawing, each tolerance interval is allocated a code, consisting of a letter and a number (Figure 8.10). The combination of reference letter and number is called the *tolerance class*.

The number indicates the *tolerance grade*, which is the magnitude of the tolerance interval. The *international tolerance (IT) grades* are standardised according ISO 286–1 and specify the difference between upper and lower deviation. However, a tolerance grade gives no information about limit sizes since it only establishes the size (thickness) of a tolerance interval, not the position with respect to the nominal size (zero line). The tolerances vary depending on nominal size ranges, but within a given IT grade the level of accuracy remains the same. Figure 8.11 depicts an overview of standard tolerance grades according to ISO 268–1. While lower numbers represent a small tolerance interval, with a larger IT grade the tolerance interval increases. In typical mechanical engineering applications we can find medium IT grades.

The letter in the ISO code system is named the *fundamental deviation* and specifies the location of the tolerance interval with respect to the nominal size. To be precise, the fundamental deviation represents the limit of size or deviation which is the nearest to the zero line (Figure 8.12).

Fundamental deviations for internal dimensions (e.g. hole) are designated by upper case letters, whereas lower case letters are used for external dimensions (e.g. shaft). Figure 8.13 depicts the fundamental deviations for internal and external dimensions according ISO 268–1. The dark grey squares represent the tolerance intervals for internal dimensions and the light grey ones for external

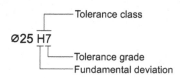

Figure 8.10 The ISO code system.

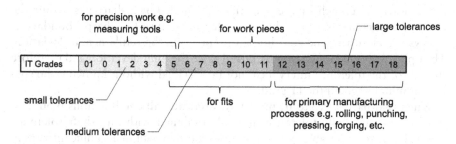

Figure 8.11 Overview of international tolerance grades.

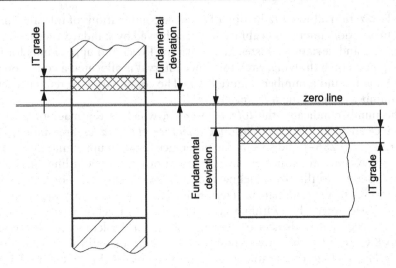

Figure 8.12 Fundamental deviation of a tolerance interval.

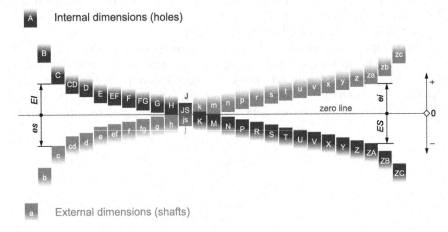

Figure 8.13 Fundamental deviations for internal and external dimensions.

dimensions. Note, each square is vertically bounded by a sharp edge while the opposite boundary is shown blurred. This is because the sharp boundary is always the closest to the zero line and represents the fundamental deviation. The opposite boundary is shown blurred since the limit deviation and therefore the size of the tolerance interval cannot be specified without knowing the specific nominal size and the IT grade.

When two parts were to be mated, theoretically all combinations would be possible, i.e. any hole tolerance interval combined with any shaft tolerance interval. The total number of combinations would result in a significant variety of machining tools and measurement devices.

8.2.4 Fit Systems

To save cost by reducing the number of combinations, two fit systems are introduced:

- *Hole-basis* fit system
- *Shaft-basis* fit system

In a hole-basis fit system, the lower size limit of the internal dimension always equals the nominal size (EI = 0). According to Figure 8.13, this leads to a fundamental deviation H for the internal mating part. The required types of fit can be obtained by combining the fixed internal feature (e.g. hole) with an external feature (e.g. shaft) of various fundamental deviations. This means that the shaft size varies according to the desired clearance or interference. As a result, all possible tolerance intervals for the external dimension can only be combined with the fundamental deviation H for the internal dimension (Figure 8.14). This leads to greater economy of production. A limited set of machining tools such as drills and reamers is sufficient to achieve a variety of fit types since external features usually can be accurately adjusted in size.

Similarly to the hole-basis fit system, the shaft-basis fit system is characterised by an upper size limit of the external dimension equal to the nominal size (es = 0). A tolerance interval which fulfils this requirement is the one with the fundamental deviation identifier h (Figure 8.15). In this fit system, the size of the internal dimension varies according to the combined tolerance interval of the external dimension. Compared to the hole-basis fit system, more drills and reamers as well as measurement devices are necessary. Therefore, this fit system tends to be more costly. Although this fit system is rarely applied, there are some applications where this system is suitable and justifies higher cost. Examples are cold-finished shafts used in textile machinery, shafts in agricultural machines, and applications in precision engineering.

Although the variety of possible combinations of tolerance intervals could be reduced by introducing the two fit systems, there are still many remaining

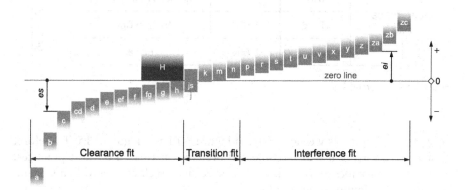

Figure 8.14 Hole-basis fit system.

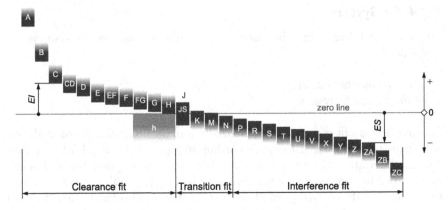

Figure 8.15 Shaft-basis fit system.

ISO CODE		DESCRIPTION
Hole Basis	Shaft Basis	
H11/c11	C11/h11	Loose Running
H9/d9	D9/h9	Free Running
H8/f7	F8/h7	Close Running
H7/g6	G7/h6	Sliding
H7/h6	H7/h6	Locational Clearance
H7/k6	K7/h6	Locational Transition
H7/n6	N7/h6	Locational Transition
H7/p6	P7/h6	Locational Interference
H7/s6	S7/h6	Medium Drive
H7/u6	U7/h6	Force

Clearance Fits — Transition Fits — Interference Fits

growing clearance — growing interference

Figure 8.16 Preferred fits.

options to combine as we can see from Figure 8.14 and Figure 8.15. Therefore, designers can make use of documented preferred fits based on best practices. Figure 8.16 provides an overview of selected fits which can serve as a starting point for determining a desired fit.

So how does one determine the type of fit if two tolerance classes are given? As an example, for two mating parts the tolerance classes H7 and g8 are given. First, we can conclude that the hole-basis fit system applies (fundamental deviation H for the internal dimension). In a second step, we need to identify the letter for the fundamental deviation of the external dimension. In this example it is g. According to Figure 8.14 this results in a clearance fit. If we would have the combination H8/k6, then this would result in a transition fit. The hole-basis fit system would still apply. As one can see, to determine whether it is a clearance, transition or interference fit, it is sufficient to check the fundamental deviations of the two mating parts. The nominal size or the IT grade is not necessary. These values are only required when we want to further specify the degree of clearance or interference, i.e. determining the upper and lower deviations.

8.2.5 Designation of Tolerances in Drawings

When a designer has specified a specific type of fit for two mating parts, there are two options to indicate in a drawing. Either the dimension consists of the nominal size followed by the designation of the required tolerance class (ISO code system), or the nominal size is followed by its signed deviations according to ISO 14405–1. In general, both methods are equivalent. The chosen concept usually depends on company internal rules. For information purposes the deviations or the tolerance class can be added in parentheses. Figure 8.17 shows different methods for the same exemplary dimension.

8.2.6 Tolerance Stack-Up

Whenever geometric features within a single part are related to each other or parts fit together in an assembly, the allowable size deviations, defined by general tolerances, specific tolerances or fits, accumulate. Whether the resulting clearance or interference is permissible can be assessed by calculating the *tolerance stack-up*. To do so, all dimensions forming a linear chain are examined regarding their permissible limits. The *closing dimension* is the one which spans the distance between the starting and end point of the dimension chain. The tolerance of the dimension chain, the so-called *closing tolerance*, is the accumulation of the individual tolerances and can be calculated by arithmetical tolerance calculation or statistical tolerance calculation which takes statistical probability into account. This books gives a very brief introduction to the arithmetical calculation.

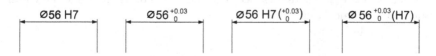

Figure 8.17 Methods of designating an internal dimension and its tolerance.

The following steps are required to arithmetically determine the tolerance stack-up:

1 Identify all dimensions which form a dimension chain.
2 Determine whether they contribute to an increase (to be drawn in positive direction) or decrease (to be drawn in negative direction) of the closing dimension.
3 Consider the worst case when maximum clearance or minimum interference is achieved. For that, sum up the upper limits of size of the positive chain links and the lower limits of size of the negative chain links. This will result in a maximum value of the closing dimension.
4 Consider the worst case when minimum clearance or maximum interference is achieved. Analogously to the previous step, sum up all lower limits of size of the positive chain links and the upper limits of size of the negative chain links. The result represents the minimum value of the closing dimension.

To determine whether a dimension contributes to an increase or decrease of the closing dimension, it can be thought of the loosest and the tightest situation. Maximum clearance (or minimum interference) is achieved when e.g. minimum sizes of physical parts and maximum sizes of gaps are considered.

To illustrate the procedure, a simple example is provided. In Figure 8.18 a typical situation is given. A deep groove ball bearing is mounted on a shaft journal, axially fixed by a retaining ring. The width of the ball bearing l_2 is 9 mm with a lower permissible deviation of –0.1 mm and an upper deviation of 0 mm. The width of the retaining ring l_3 is 1 mm. The lower permissible deviation is –0.05 mm and the upper deviation 0 mm. A lateral clearance s of 0 mm up to maximum 0.2 mm is accepted. What is dimension l_1 and its tolerance?

As previously explained, it is essential to consider the worst cases only. In general, the overall length l_1 is calculated as

$$l_1 = l_2 + l_3 + s$$

The first case to consider is the loosest scenario, i.e. defining the maximum closing dimension.

Case 1 (loose):

$$l_{1,max} = l_{2,min} + l_{3,min} + s_{max}$$
$$\Rightarrow l_{1,max} = 8.9 + 0.95 + 0.2 = 10.05\,mm$$

In the second case the tightest scenario is examined, which leads to the minimum closing dimension.

Case 2 (tight):

$$l_{1,min} = l_{2,max} + l_{3,max} + s_{min}$$
$$\Rightarrow l_{1,min} = 9.0 + 1.0 + 0 = 10.0\,mm$$

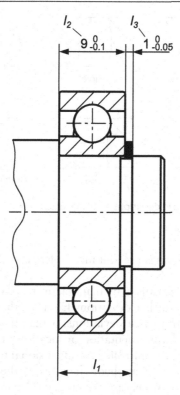

Figure 8.18 Example of tolerance stack-up.

According to the calculation, the lower limit of size of the closing dimension is 10.0 mm and the upper limit of size is 10.05 mm. Therefore, $l_1 = 10 +0.05$.

As a rule, arithmetical tolerance calculation leads to tolerances which are too tight and therefore contribute to higher cost and increased effort for inspection. For that reason, statistical methods are used in industry. However, due to the complexity of this topic, a comprehensive introduction to tolerance management would be out of the book's scope.

8.3 GEOMETRICAL TOLERANCING

8.3.1 Introduction to Geometrical Tolerancing

Engineering drawings are supposed to show a part with all necessary details in such a way that it can be reproduced. Therefore, the representation on the drawing and the complementary indications need to be complete and without ambiguity. Any drawing which does not fulfil these requirements is not suitable for production and quality assurance (QA). In past decades, raised expectations regarding improved part quality led to higher requirements. In fact, most of

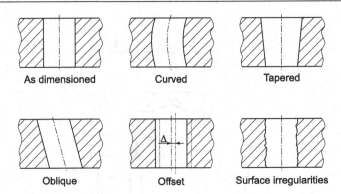

Figure 8.19 Possible geometric deviations after manufacturing.

today's engineering drawings are still incomplete and ambiguous due to missing specifications.

While deviations in size caused by manufacturing issues can be controlled by dimensional tolerances, further deviations from the theoretical (perfect) geometry can occur. Apart from size deviations, a manufactured part can deviate regarding the intended form, orientation or location of its features. These geometric deviations cannot be controlled by dimensional tolerancing. Instead, geometrical tolerances have to be applied. They specify the allowable variations in form or position to ensure proper functioning of the part and interchangeability.

Geometric imperfections can have multiple causes such as operating errors of machines, inappropriate tool fixtures, machine vibrations, worn out tools and environmental conditions (e.g. temperature, humidity). Figure 8.19 provides an overview of possible geometric deviations, exemplarily shown for a drilled hole.

The hole can deviate in form, for example due to too much force applied onto the drill or to worn out cutting edges of the drill bit. If the driller is not oriented perpendicularly to the part's top surface, the resulting hole can be oblique. Maladjustment of the tool can lead to an offset. The hole will not be cylindrical and of smooth surface if there are machine vibrations or worn out cutting edges.

Figure 8.20 illustrates an example of a simple plate, shown with all dimensions and tolerances as needed for manufacturing. For clarity, the drawing does not include any surface texture requirements or additional annotations and general tolerances are explicitly shown with the dimensions. In Figure 8.21 the same plate is shown but with actual sizes. Note, the resulting deviations are shown exaggerated for clarity.

Obviously the plate does not finally look as initially given on the drawing. However, all deviations are permissible. Here are some examples:

–   The size 94.32 mm for the overall actual width is permissible since the given dimension for the width 92±0.3 refers to the distance between the two lateral edges, which actually is 91.93 mm. The direction of

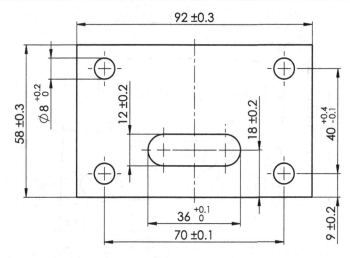

Figure 8.20 Drawing of a plate including dimensions and tolerances.

measurement was not specified, therefore the direct distance between the edges is taken.

- The actual vertical position of the slotted hole is permissible. If measured from the bottom right corner of the plate, the vertical position would be out of the allowable range. Since the dimension 18±0.2 refers to the bottom edge and not any corner, the measured actual size of 18.15 mm is within the range of limits.
- The actual horizontal distance of 70.07 mm (82.47 mm – 12.4 mm) between the holes is permissible since the allowable distance was specified with 70±0.1. The two holes are offset to the left by more than 1 mm, but this is acceptable due to missing tolerance.
- The angular offset of 2.3° for the lateral edges is permissible since there is no tolerance indicated regarding the angularity.

It can be assumed that the function of the part would not be given with the actual dimensions shown in Figure 8.21 although all dimensional specifications have been fulfilled. To avoid possible problems during assembly or operation, the specifications regarding form, location and orientation need to be more precise. Geometrical tolerances help to achieve geometrical precision.

According to ISO 1101, geometrical tolerances can be classified as shown in Figure 8.22.

8.3.2 Datums

For some geometrical tolerances, it is essential to establish a datum which can be seen as a reference. The theoretical datum system can consist of a datum point, datum planes which for example are considered to be the origin for

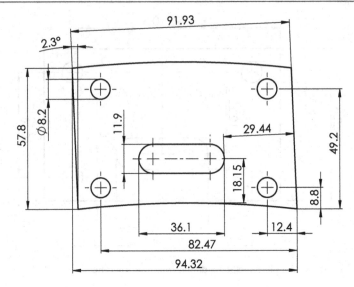

Figure 8.21 Drawing of a plate with actual sizes after manufacturing.

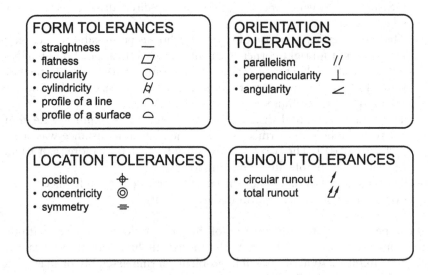

Figure 8.22 Classification of geometrical tolerances according ISO 1101.

measurements, and datum lines or axes. Datums also contribute to the definition of a tolerance zone.

If a toleranced feature refers to another entity such as an edge, surface or centre line/plane, then this entity serves as a *datum feature* for the geometrical tolerance. Note, a datum feature is not equal to the datum of the geometrical tolerance. A datum is a geometrically ideal entity such as a point, line, surface or plane. It is therefore an abstract geometric concept and only theoretical. In

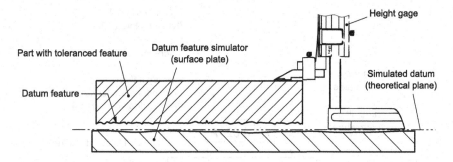

Figure 8.23 Datum feature and simulated datum.

contradiction, a datum feature is a functional entity of a part, which means it is tangible. Due to manufacturing related deviations from the perfect geometry, a datum feature will never be identical to the ideal datum which is simulated by measuring equipment.

To visualise the importance of datums and the simulation of datum features, Figure 8.23 depicts how a part's height is measured with a height gauge on a surface plate.

In this example, a part is geometrically toleranced and for quality assurance the height of the part is measured with a height gauge. The datum feature is the bottom face of the part. Owing to manufacturing, the bottom face shows some imperfections and is not geometrically ideal. The height gauge is placed on a surface plate which serves as the simulated datum. It is considered to be the origin of measurement. Since the existing imperfections on the top face of the surface plate will be far smaller than those of the datum feature, the accuracy of measurement will be given. However, there will always be a remaining uncertainty of measurement due to the difference between the actual surface of the surface plate and the theoretically ideal datum plane.

8.3.3 General Rules for Indication

A geometrical tolerance is indicated on a technical drawing by using a graphical symbol. This so-called *feature control frame* usually consists of three sections: the geometrical tolerance symbol, a value for specifying the tolerance zone and a letter referencing a datum (if applicable), compare Figure 8.24.

The leader line with the arrowhead points to the feature the geometrical tolerance is applied to. Alternatively, it can point to the corresponding extension line. In the feature control frame, the tolerance value is preceded by the diameter symbol ($\varnothing$) if the tolerance zone is diametrical. The tolerance value is given in the same units as the dimensions on the drawing. In the last section of the feature control frame, one or more capital letters are used to indicate a datum the tolerance refers to. Note, not all geometrical tolerances require a datum. If two datums are required, the reference letters are shown with a hyphen in between. If more than one datum is indicated and a certain sequence shall be

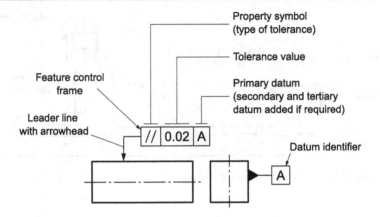

Figure 8.24 Feature control frame and datum identifier.

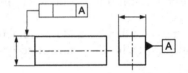

Figure 8.25 Surface as toleranced feature.

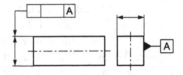

Figure 8.26 Midplane as toleranced feature.

kept, the datum letters are separately boxed, with the datum with the highest priority mentioned first.

The datum feature is denoted on a drawing with a datum identifier consisting of a box, which includes the reference letter and a black triangle the box is tied to. The black triangle is placed on the datum feature.

Special attention is needed when placing the feature control frame. If the toleranced feature is a surface or line, the leader line of the feature control frame can be placed either on the surface/line itself or on an attached extension line clearly apart from any dimension line (Figure 8.25).

However, if the tolerance refers to a centre line or midplane, the leader line must be aligned with the corresponding dimension line. The same rules apply to the placement of a datum identifier. Figure 8.26 depicts an example where the midplane of the prismatic part constitutes the toleranced feature. The black triangle of the datum identifier is also placed apart from any dimension line, therefore it refers to the lateral face of the part.

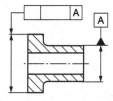

Figure 8.27 Centre line as datum feature.

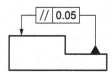

Figure 8.28 Feature control frame directly connected to the datum feature.

Figure 8.27 illustrates an example of a rotationally symmetric part, the axis of the largest cylindrical face of which is the toleranced feature. The axis of the smaller cylinder is the datum feature since the datum identifier is aligned with the dimension line of that cylinder.

The complete symbol for the datum identifier is not needed if the feature control frame can be directly connected to the datum feature without ambiguity (Figure 8.28).

8.3.4 Types of Geometric Specification

Form Specifications

Whenever accuracy in shape of a feature is necessary, it can be controlled with form tolerances. Each type of form tolerance has its individual definition of a tolerance zone within which all points of the toleranced feature must be contained. Any form of the feature is permissible, as long as it is within the specified tolerance zone.

The form tolerance **straightness** can be applied to any linear feature that is a straight line or a set of straight lines. Figure 8.29 shows an example for a set of straight lines. Any extracted line from the indicated face of the part shall be contained within the given tolerance zone. In other words, at any position across width b, the profile line must be within two parallel lines spaced $t = 0.08$ mm apart.

Alternatively to a set of straight lines, the straightness tolerance can be applied to a single line. Figure 8.30 illustrates how the axis of a cylindrical part is toleranced. In this case the tolerance zone is a cylinder of $t = 0.05$ mm diameter.

Typically, a straightness tolerance is used for sealing surfaces, i.e. when two parts mate each other and form a seal. Straightness can also be applied to

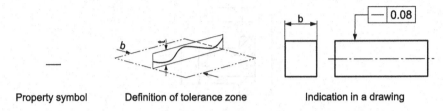

| Property symbol | Definition of tolerance zone | Indication in a drawing |

Figure 8.29 Geometrical tolerance for straightness of a set of lines.

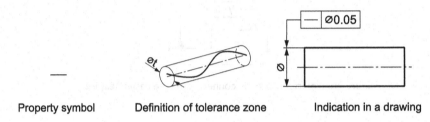

| Property symbol | Definition of tolerance zone | Indication in a drawing |

Figure 8.30 Geometrical tolerance for straightness of a single line.

| Property symbol | Definition of tolerance zone | Indication in a drawing |

Figure 8.31 Geometrical tolerance for flatness.

cylindrical surfaces which are assembled into a bore or hole and require a certain clearance, or when parts are prone to bending.

The form tolerance **flatness** can be applied to planar faces only. The extracted surface shall be contained between two parallel planes which are at a distance of the tolerance value t. The toleranced feature is considered to be flat as long as each point of the surface is within the tolerance zone. Figure 8.31 depicts an example.

Flatness is measured with a height gauge running across the toleranced surface. For that, it is essential to make sure that the reference for the gauge is in parallel. Nowadays, coordinate measuring machines (CMM) are able to compare the measured surface with a virtual plane. A flatness specification is used, for example, for surfaces forming a seal with a mated part or for ensuring a uniform wear of a surface.

Circularity, also known as *roundness*, is used for specifying the form of a circular line or a set of circular lines. This can be applied to either cylindrical features, conical features, spheres, or any other rotationally symmetric feature.

The circumferential line in any cross-section of the toleranced surface must be within two coplanar concentric circles with a radial distance of the tolerance value t. In case of cylinders, the cross-sections are oriented perpendicular to the axis of the cylinder. It can be assumed that this also applies to conical surfaces. However, to avoid misunderstanding it is recommended to explicitly specify the orientation by adding a direction feature indicator. For spheres, the cross-section shall contain the centre point. In Figure 8.32 an example for a cylindrical and a conical surface is shown.

Similar to a circularity tolerance, **cylindricity** limits the deviation in the radial direction. This kind of geometrical tolerance can be applied to surfaces only. The tolerance zone is a shell consisting of two coaxial cylinders with a radial distance of the tolerance value t. In the example shown in Figure 8.33, the toleranced cylindrical surface shall be within two coaxial cylinders 0.05 mm apart.

The form tolerance **profile of a line** controls the form of a non-straight single line or set of lines. The line of interest needs to be dimensioned with theoretical exact dimensions. In each cross-section parallel to the plane of projection where the geometrical tolerance is indicated, the profile line whose centre points are positioned on a line representing the geometrical ideal form shall be within two equidistant lines enveloping circles of tolerance diameter. In the example shown in Figure 8.34, the profile line consisting of two straight lines and two coalescing arcs must be within two equidistant lines 0.02 mm apart.

The profile of a line can be measured by using either a gauge or, for complex geometries, a coordinate measuring machine. A form specification of this type can be used for surfaces of multiple curvature. Examples are complex freeform

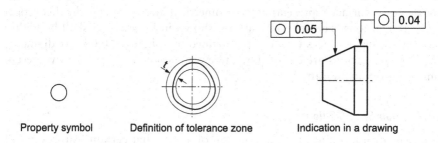

Property symbol   Definition of tolerance zone   Indication in a drawing

Figure 8.32 Geometrical tolerance for circularity (roundness).

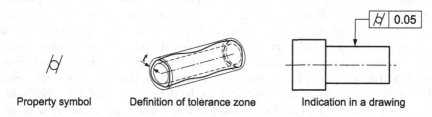

Property symbol   Definition of tolerance zone   Indication in a drawing

Figure 8.33 Geometrical tolerance for cylindricity.

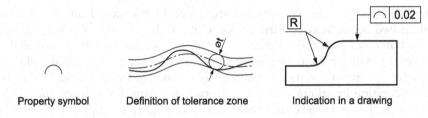

Property symbol   Definition of tolerance zone   Indication in a drawing

Figure 8.34 Geometrical tolerance for the profile of a line.

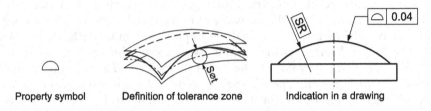

Property symbol   Definition of tolerance zone   Indication in a drawing

Figure 8.35 Geometrical tolerance for the profile of a surface.

geometries as they can be found in the automotive industry, in aviation and shipping.

While the line profile tolerance is for linear features, the **profile of a surface** tolerance is for areal features. The definition of the tolerance zone is similar to that of the line profile, except for the third dimension which is now taken into account. The permissible deviation of the profile of a surface is limited by two surfaces enveloping spheres of tolerance diameter, the centres of which are located on a surface representing the geometrical ideal form, i.e. the theoretical exact dimensions. The spherical surface shown in Figure 8.35 shall be within two enveloping surfaces whose gap is defined by spheres of 0.04 mm diameter. Typical applications are casted parts, freeform geometry in automotive design and wings of an aircraft.

Orientation Specifications

It can be of importance for the function of a part that certain features are of well-defined orientation. A feature's orientation can be controlled with orientation tolerances. This kind of geometrical tolerance requires a minimum one datum. As an example, it is impossible to specify that a feature shall be parallel or perpendicular if there is no reference. For that reason, orientation tolerances are also designated as *related geometrical tolerances*. Each point of a toleranced feature must lie within the specified tolerance zone. As a rule, orientation tolerances specify the permissible deviation from the geometrical ideal orientation, given by one or more datum features. At the same time they limit the form deviation of the toleranced feature due to the definition of the tolerance zone. Note, this does not apply to the datum feature(s). Therefore, it might be necessary to add a form tolerance to the datum feature(s).

The orientation tolerance **parallelism** can be applied to either linear or areal features. This geometrical tolerance controls the orientation of a line or surface with respect to a given reference. When tolerancing a surface, the feature shall be contained between two parallel planes whose distance is given by a tolerance value. The two planes defining the tolerance zone need to be parallel to a given datum. In the example shown in Figure 8.36, the indicated flat surface shall be within a tolerance zone of 0.02 mm thickness in which planes are parallel to the bottom surface designated as datum A.

Note, with a parallelism tolerance it is not possible to control the angle of the toleranced surface. Only the envelope of the surface is under control. As an example, parallelism specifications are sometimes applied to lateral faces of gears to limit their deviation in width.

Figure 8.37 illustrates how the tolerance zone is defined for a linear feature and provides an example how to indicate in a drawing. The axis of the smaller hole shall be within a cylindrical tolerance zone of diameter 0.02 mm, the centre line of which is parallel to the axis of the larger hole, given as datum A.

When two faces, either linear or areal, need to be at right angle then the geometrical tolerance **perpendicularity** can be applied. A typical example is a flange bearing. When tolerancing a surface, the feature shall be contained between two parallel planes which distance is given by a tolerance value. The two planes defining the tolerance zone need to be perpendicular to a given datum. In the example shown in Figure 8.38 the indicated lateral surface whose planes are perpendicular to the hole's axis designated as datum A shall be within a tolerance zone of 0.1 mm thickness.

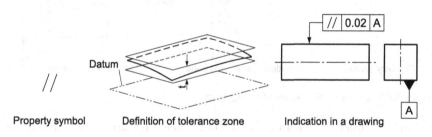

Figure 8.36 Geometrical tolerance for parallelism of a surface.

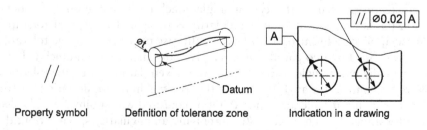

Figure 8.37 Geometrical tolerance for parallelism of a line.

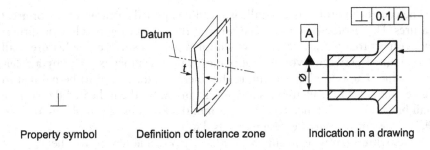

| Property symbol | Definition of tolerance zone | Indication in a drawing |

Figure 8.38 Geometrical tolerance for perpendicularity of a surface.

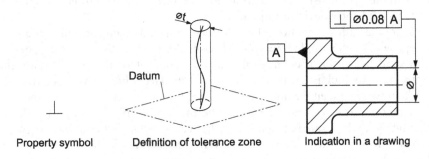

| Property symbol | Definition of tolerance zone | Indication in a drawing |

Figure 8.39 Geometrical tolerance for perpendicularity of a line.

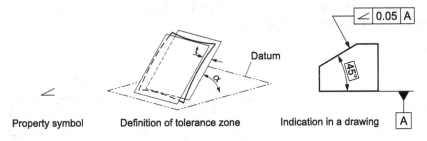

| Property symbol | Definition of tolerance zone | Indication in a drawing |

Figure 8.40 Geometrical tolerance for angularity.

Since the toleranced feature can also be a linear feature such as an edge or a centre line, the corresponding datum must be of planar type. Figure 8.39 gives an example. The centre line of the hole must be within a cylinder of diameter 0.08 mm that is perpendicular to the plane face designated as datum A.

For features oriented at a certain angle which need precise limits for deviation, the geometrical tolerance **angularity** is of help. This type of tolerance can be applied to linear or areal features. In case of plane faces, the tolerance zone is defined by two parallel planes a given distance apart and inclined at a theoretically exact angle with reference to a given datum. Note, the tolerance indirectly controls the angle because of the specified limits for deviation. Therefore, the angularity tolerance should not be confused with a dimensional tolerance for an angle such as ±2°. When applying an angularity specification, the angle dimension becomes a theoretically exact dimension. In Figure 8.40 the indicated plane surface shall be within two parallel planes 0.05 mm apart and at a theoretical angle of 45° with reference to the bottom surface (datum A).

Location Specifications

For proper functioning of a part it might be essential to further specify the location of a feature. This especially applies to theoretical features such as centre lines or midplanes which cannot be directly measured. Points of a toleranced feature shall be within the specified tolerance zone, which for these kinds of geometrical tolerances is always assumed to be in the geometrical ideal orientation. For that reason, location tolerances are also related geometrical tolerances and therefore require one or more datums. The presented location tolerances specify the permissible deviations from its geometrical ideal location. A feature's location can be specified by its orientation and its distance from any given datum. Therefore, by tolerancing the location of a feature, at the same time the feature's orientation and form is controlled. Note, the form deviation of the datum feature(s) is not limited by location tolerances. For better control it might be necessary to add a form tolerance to the datum feature(s).

When using the geometrical characteristic symbol for **position,** the tolerance feature can be either a point, a straight line such as an edge or an axis, or a planar face. The location of the toleranced feature shall be dimensioned with theoretical exact dimensions only. Figure 8.41 depicts an example where the position of the centre line of a hole is toleranced. The centre line shall be contained within a cylinder of diameter 0.06 mm, having its centre at the true location 5 mm and 6 mm apart, respectively, from the planar faces, designated as datum A and datum B. In addition, the cylinder must be perpendicular to datum C.

The location tolerance **concentricity** refers to points, a set of points or a straight line. It can be used to control the maximum deviation of concentricity of two circular shapes. Per definition, two circles are concentric when their centres are coincident. This is given when the centre point of the toleranced circular shape is within a well-defined circle, the centre point of which coincides with a given datum point. According to Figure 8.42, in any cross-section the deviation of the centre point of the inner circle from the true centre is controlled by the

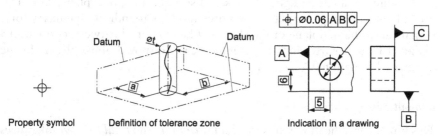

Figure 8.41 Geometrical tolerance for the position of a line.

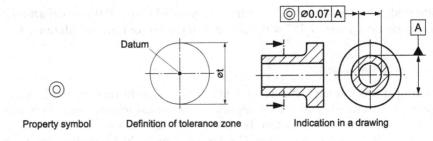

Property symbol   Definition of tolerance zone   Indication in a drawing

Figure 8.42 Geometrical tolerance for concentricity.

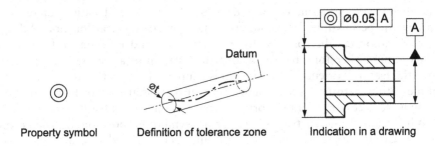

Property symbol   Definition of tolerance zone   Indication in a drawing

Figure 8.43 Geometrical tolerance for coaxiality.

circular tolerance zone of 0.07 mm diameter. The true centre in this case is the centre point of the larger circle in the given cross-section.

When cylindrical features are expected to be aligned and therefore require accurate positioning relative to each other, a **coaxiality** tolerance can be applied. It is defined as analogous to concentricity, except that the third dimension is taken into account. For that reason, the property symbol is the same. When tolerancing with regard to coaxiality, the tolerance zone is a cylinder with diameter t, the axis of which coincides with a given linear datum. In Figure 8.43 the centre line of the larger cylindrical feature is supposed to be within a cylinder of diameter 0.05 mm, whereas the axis of the cylindrical tolerance zone is coincident with the centre line of the smaller cylindrical feature.

For tolerancing symmetrical features, the geometrical tolerance **symmetry** is the right choice. This type of location tolerance can be applied to points, a set of points, a single straight line, a set of straight lines, or a planar face. Figure 8.44 exemplarily shows the definition and how to indicate symmetry for a planar face. The midplane of the slot shall be contained between two parallel planes 0.1 mm apart which are symmetrically disposed about the midplane designated as datum A.

Run-out Specifications

This kind of specification refers to *circular run-out* and *total run-out* tolerances in axial or radial direction. Although these geometrical tolerances represent a

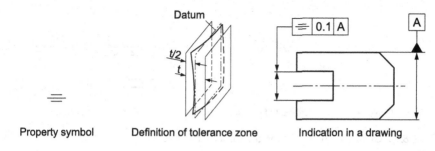

Property symbol   Definition of tolerance zone   Indication in a drawing

Figure 8.44 Geometrical tolerance for symmetry.

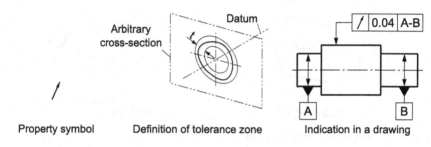

Property symbol   Definition of tolerance zone   Indication in a drawing

Figure 8.45 Geometric tolerance for circular radial run-out.

special type of orientation and location tolerances, according to ISO 1101 they are considered to be a separate category due to the required measuring method which differs from the other geometrical tolerances.

The tolerance type **circular radial run-out** is for limiting the radial deviation of a circular line or a set of circular lines. The profile line in any cross-section of the toleranced feature shall be contained between two coplanar concentric circles of a given radial distance *t*. The cross-section planes are supposed to be perpendicular to a given datum axis. In practice, a radial run-out is measured with the help of a height gauge. The part is constrained with V-blocks and rotated 360°. During rotation, the radial variation is measured with the gauge, held perpendicular to the toleranced surface. Figure 8.45 depicts an example of a simple shaft. In each cross-section of the cylindrical feature, the deviation of the extracted circumferential line is limited by two coplanar circles 0.04 mm apart, the centre of which is concentric to the common datum axis A–B. Each cross-section is perpendicular to the datum axis.

Typical applications for the use of radial run-outs are shafts, spindles, rotary tables and rotary stages. For a rotary table, radial run-out can affect the function which is to centre a part on the table. Exceeding given limits can lead to *eccentricity*.

Another type of run-out tolerance is the **circular axial run-out**. Axial run-out defines the motion of a flat surface in axial direction when it rotates. This can be of special interest in applications such as metrology or inspection. Per

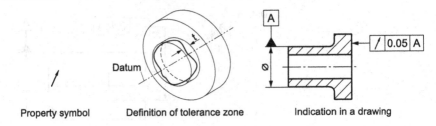

Property symbol   Definition of tolerance zone   Indication in a drawing

Figure 8.46 Geometric tolerance for circular axial run-out.

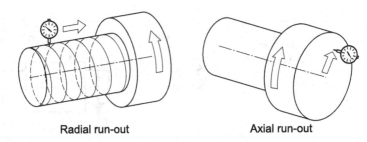

Radial run-out   Axial run-out

Figure 8.47 Measurement of total run-outs.

definition, the circumferential line in each cylindrical section coaxial with a given datum axis shall be contained between two circles at distance t and perpendicular to the datum axis. Figure 8.46 depicts an example. In every section, the extracted line is limited by two circles 0.05 mm apart. The axis of the cylindrical section coincides with the axis of the referenced cylindrical feature, designated as datum A.

In contrast to circular run-out, a total run-out is defined as the deviation measured across the entire surface rather than at a single point on the toleranced surface. The toleranced feature can be a cylindrical surface or a flat surface. A total run-out specification controls the permissible variation on the surface as the part is rotated 360° around the datum axis, and the variation in axial direction as well. Total run-out specifications can be used to prevent vibration and oscillation. Typically, total run-out tolerances are applied to transmission shafts or complex gears which require precision. Figure 8.47 illustrates how total run-out is measured.

In case of **total radial run-out** specification, the toleranced cylindrical surface shall be contained between two coaxial cylinders differing in size by a given tolerance t. The axes of the cylinders coincide with a given datum which usually is an axis.

For **total axial run-out**, the toleranced flat surface must be between two parallel planes spaced apart at a given distance of tolerance value t and perpendicular to a given datum line. Figure 8.49 illustrates an example. The right-hand lateral face of the given part shall be contained between two parallel planes

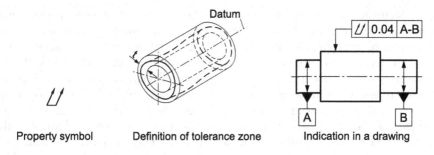

Property symbol   Definition of tolerance zone   Indication in a drawing

Figure 8.48 Geometric tolerance for total radial run-out.

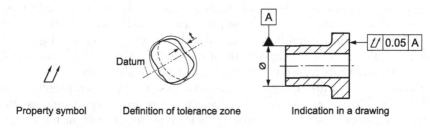

Property symbol   Definition of tolerance zone   Indication in a drawing

Figure 8.49 Geometric tolerance for total axial run-out.

0.05 mm apart and perpendicular to the axis of the left-hand outer cylindrical face.

8.4 SURFACE TEXTURE

8.4.1 Surface Roughness Parameters

As a rule, a part consists of a variety of functional surfaces and might be exposed to different types of mechanical stress. Therefore, apart from material properties, the functionality of a part depends on its geometrical properties. Functional surfaces can be classified into three groups:

- Surfaces with no or low mechanical stress
- Surfaces exposed to mechanical stress but without relative movement to opposing surfaces
- Surfaces exposed to mechanical stress and with relative movement to opposing surfaces

In engineering drawings all surfaces are considered to be geometrically perfect. In fact, the resulting surfaces are characterised by more or fewer imperfections due to, for example, human mistakes, worn out tools and vibrating machines. Sometimes the surface quality is simply a question of manufacturing cost. As a rule, higher accuracy requires specialised tools and/or more time for manufacturing. In contrast to the theoretically perfect *geometric surfaces*, we obtain *real surfaces* which are imperfect and accompanied by form deviations. Examples of imperfections are corrugations, scratches, burrs and cracks.

Form deviations can be classified into deviations of different orders. While deviations of first order describe simple shape deviations such as deviations from flatness, roundness or straightness, deviations of higher order refer to smaller deviations. Deviations of second order describe the *waviness* of a real surface, i.e. periodically occurring deviations. Deviations of third and fourth order characterise the *roughness* of a face and finally deviations of fifth and sixth order refer to imperfections in the material structure and are not visible without microscopy.

While deviations of first and second order can be controlled with geometrical tolerancing, deviations of third and fourth order can only be controlled by indicating a desired surface roughness on the drawing. Surfaces frequently require a well-defined surface finish, for example where parts fit together tightly or where relative movement between them is required. When parts form a seal, special requirements regarding the surface texture are also given.

Deviations from first to fourth order superpose to the actual profile, or in other words, when measuring the real surface, it is necessary to suppress the longwave component of the primary profile (*P-profile*) to obtain the roughness profile, also designated as *R-profile*. Figure 8.50 depicts a piece of surface, showing the actual profile. From this it is possible to plot the primary profile and to determine the corresponding waviness profile (*W-profile*) as well as the roughness profile.

A typical plot of a primary profile is shown in Figure 8.51. The mean line represents the average level of the surface. According to ISO 4287 and ISO 4288, the individual profile peak heights Z_p and profile valley depths Z_v can

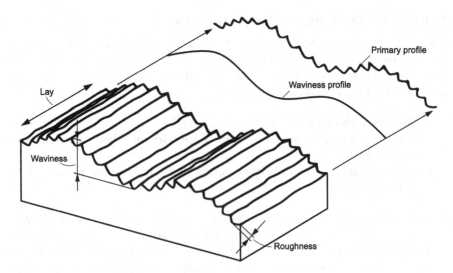

Figure 8.50 Actual profile of a surface.

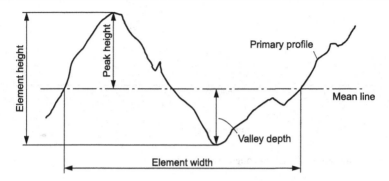

Figure 8.51 Profile element of a primary profile from an actual surface.

be determined. The sum of the peak height and valley depth of one profile element is designated as profile element height Z_t. The 'period' of the wave is measured as the profile element width X_s.

When the roughness of a real surface is measured, it is not sufficient to consider just one profile element. Instead, the primary profile of a well-defined sampling length is determined in such a way that several profile elements are part of the measurement. As a result we obtain maximum occurring values for upper introduced parameters of a profile element.

With time, 65 different surface roughness parameters were developed, differing in the applied filter to separate W-profile and R-profile. Most commonly used one-dimensional roughness parameters are:

- Arithmetic mean of profile ordinates, *Ra*
- Average peak to valley height, *Rz*

While the definitions for *Ra* and *Rz* are standardised according ISO 4287 so that there is a common understanding in industry, the designation frequently differs. For that reason you might find a term such as 'average roughness' for *Ra* or 'mean roughness depth' for *Rz*. The roughness parameter *Ra* is widely used, often for historical reasons. This parameter is particularly used in the United States. Meanwhile, the parameter is often replaced by *Rz*, mainly due to some drawbacks of *Ra*.

The surface roughness parameter *Ra* is defined as the arithmetic mean of the absolute ordinate values *Z(x)* of the R-profile within the sampling length *lr*:

$$Ra = \frac{1}{lr} \times \int_0^{lr} |Z(x)| \, dx$$

Figure 8.52 depicts the definition of surface roughness parameter *Ra*. Note, *Z(x)* denotes the ordinate value which is the height of the profile at any position *x*. The rectangular area defined by the *Ra* value and the sampling length *lr* is identical to the shaded area, which indicates the total deviation from the

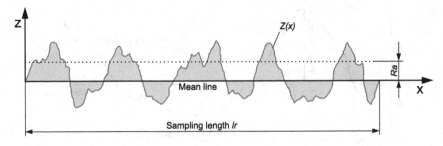

Figure 8.52 Definition of surface roughness parameter Ra.

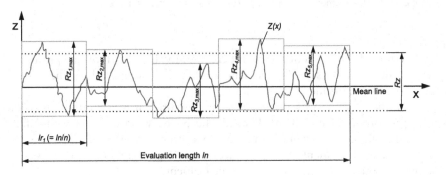

Figure 8.53 Definition of surface roughness parameter Rz.

mean line. Since Ra represents the average variation, the Ra value is lower than the maximum peak height of the R-profile and therefore this parameter is not sensitive to spike-like peaks and valleys. Owing to this drawback, it is possible to obtain the same Ra value for work pieces with different surface textures. Although Ra is not suitable, this parameter is still widely in use.

The surface roughness parameter Rz eliminates the drawback of Ra or at least reduces the averaging effect in case of high peaks. The parameter Rz is more sensitive to spikes.

To calculate Rz, the evaluation length of the R-profile is divided into n (usually 5) segments of similar lengths lr. In a second step the maximum profile heights $Rz_{1,max} - Rz_{5,max}$ are determined for each individual segment. A maximum profile height within a single sampling length lr can be obtained as the distance between the highest peak and the deepest valley. The value Rz is finally calculated as the arithmetic average of the n maximum profile height values:

$$Rz = \frac{1}{n}\sum_{i=1}^{n} Rz_{i,max}$$

The definition of surface roughness parameter Rz is visualised in Figure 8.53. Owing to the calculation method for Rz, the more spikes occur, the greater the

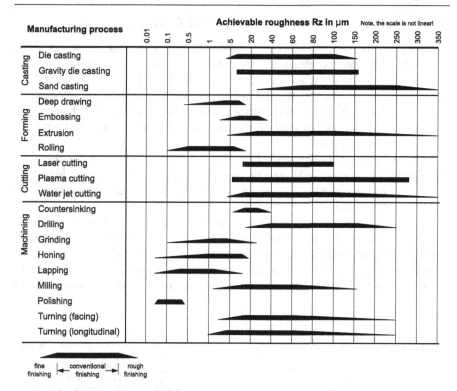

Figure 8.54 Achievable surface qualities by different manufacturing processes.

impact on the *Rz* value. In contrast to *Ra*, the parameter *Rz* is based on maximum values, therefore the values for *Rz* are usually greater compared to *Ra*.

The two surface roughness parameters *Ra* and *Rz* cannot be easily converted from one to the other since there is no mathematical relationship between their definitions. Instead, available conversion charts or conversion factors for rough estimation can be used. The more uniform a profile of a real surface is, the more both parameter values will get closer to each other.

Analogous to dimensional tolerances, the requirements for surface quality have an exponential effect on manufacturing cost. The higher the desired surface accuracy, the more additional manufacturing steps are required which leads to higher cost. In addition, efforts for quality control increase. Therefore, the same principle applies as for tolerances and fits: *As rough as possible, as fine as required for the function of the part.*

Manufacturing processes usually allow a certain degree of control over the surface finish. As an example, when turning, smoother surface finishes can be achieved by reducing the cutting speed, feed rate and/or depth of cut. The chart shown in Figure 8.54 shows the range of achievable surface roughness for some selected manufacturing processes.

8.4.2 Indication of Surface Texture in Drawings

Surface texture specifications are indicated in engineering drawings using graphical symbols, accompanied by textual indications. According to ISO 1302, surface texture requirements are specified in drawings with the help of variants of a basic graphical symbol, supplemented with complementary values, symbols and text which further specify the requirements. Figure 8.55 shows the basic graphical symbol, the meaning of their variants and where to position additional marks.

In most drawings the expanded graphical symbol can be found, either with the closed triangle to indicate that removal of material (e.g. machining) is required, or with open triangle which means that the manufacturing process is not further specified.

If the surface texture requirement applies to all adjacent faces of the referred surface in a given closed outline, this can be indicated by adding a small circle on top of the graphical symbol, as illustrated in Figure 8.56.

In this example, just one graphical symbol is placed on a surface, denoted by 'a' in the trimetric view of the given part. The surface texture requirement Rz 16

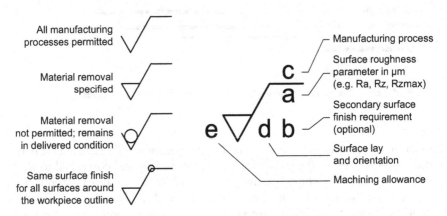

Figure 8.55 Graphical symbols for indicating surface texture requirements.

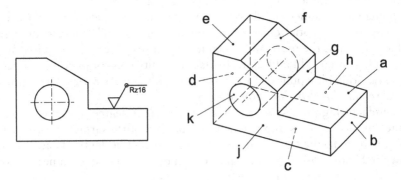

Figure 8.56 Indication of surface texture requirements for adjacent faces.

refers to all adjacent faces which belong to the closed outline. That means, faces 'a' to 'g' have the same surface finish but not faces 'h' to 'k'. The face 'h' is the rear surface and 'j' is the front surface. Face 'k' is the cylindrical face of the hole. These faces do not belong to the closed outline the graphical symbols refers to.

Many companies indicate the surface roughness parameter only. However, this can vary and depends on a company's culture. Whenever a roughness parameter is indicated, it is required to indicate the abbreviation for the parameter, e.g. Ra or Rz, followed by the value in micrometres. When using English units instead of SI units, the surface roughness shall be given in micro-inches. Complementary surface texture requirements can be added to the required surface roughness. As an example, the manufacturing method or special treatment and coatings can be explicitly given above the graphical symbol. An additional machining allowance is given in mm and placed on the left-hand side of the graphical symbol.

In most material removing manufacturing processes a certain pattern of surface irregularities can be observed. With the so-called *lay* it is possible to describe this dominant pattern and to indicate its main orientation. Figure 8.57 specifies the possible directions of surface lay and which graphical symbol to use for indication in a drawing.

Note, the indicated surface lay directions shown at the top of Figure 8.57 refer to the view's plane of projection where the symbol is shown. The lay directions 'crossed' and 'multi-directional' ensure that there is no single dominant pattern direction, which for example is not recommended in case of surfaces forming a seal.

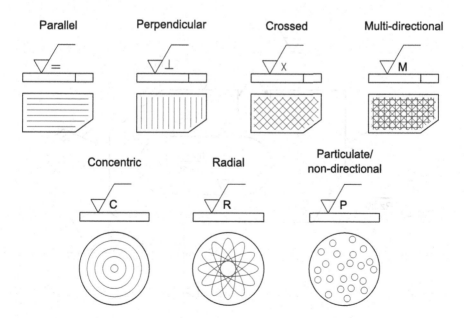

Figure 8.57 Surface lay directions and the respective symbols.

Whenever graphical symbols for surface texture contain textual information, the symbol shall be oriented in accordance with the applied reading direction (ISO 129–1). That means, if aligned dimensioning is used, the graphical symbols for surface texture and their annotations shall be readable from the bottom or right-hand side of the drawing. The symbol shall be placed in such a way that it touches the surface it refers to from outside the material. It can be placed either on a feature's outline, on an extension line, or on a reference line, pointing to the surface with a leader line. When using a leader line, remember the general rule that leader line termination is a filled dot when pointing to a surface and a filled arrowhead when pointing to an edge. Figure 8.58 depicts possible drawing entries.

If the same surface texture requirements are given for two opposite surfaces, the graphical symbol can be placed on the dimension line of the respective dimension if there is no risk of ambiguity.

Surfaces of cylindrical or prismatic features are specified only once, i.e. just on one side of the centre line.

When space is limited, surface texture symbols can be used instead of the expanded graphical symbol with all accompanying annotations. Usually the simplified symbols are explained near to the title block. Alternatively, the simplified reference indication can be explained near the view where it is used. Figure 8.59 shows examples of drawing entries.

If the majority of the surfaces requires the same surface texture, the graphical symbol for this requirement can be placed close to the title block. This

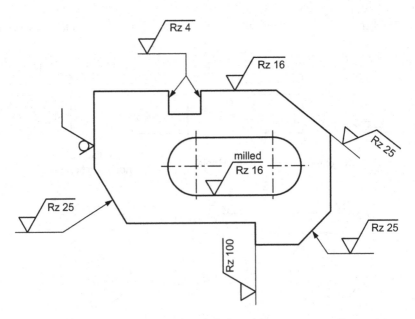

Figure 8.58 Placement and orientation of symbols for surface texture requirements.

symbol is followed by special deviating surface texture requirements, all shown in parentheses (Figure 8.60). That means, the surface finish in front of the parentheses refers to all surfaces of the part, while the surface finishes shown in parentheses represent the exceptions. The latter need to be explicitly marked in the drawing.

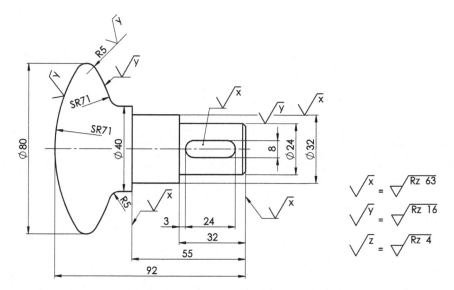

Figure 8.59 Use of simplified surface texture symbols.

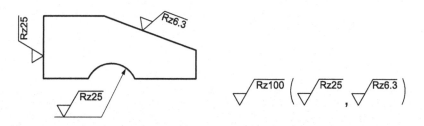

Figure 8.60 Indication of surface texture requirements for the majority of surfaces.

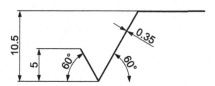

Figure 8.61 Proportions and dimensions of graphical symbols for surface texture requirements.

The graphical symbol is drawn with the same line thickness as standard lettering. Sine we usually use line group 0.5, the thickness of the graphical symbol is 0.35 mm. Figure 8.61 shows the proportions and dimensions of a basic graphical symbol for line group 0.5 according to ISO 1302.

Chapter 9

Representation of Workpiece Elements

Many designs comprise standardised machine elements to fulfil the desired function. These elements can contribute to power transmission, provide a certain support for other components of an assembly, or are used as fasteners, i.e. they hold components in place. Some of these machine elements require a modification of workpieces to be used. These modifications are sometimes referred to as workpiece elements. Introducing all machine elements would be out of the scope of this book. However, this chapter presents a selection of commonly used workpiece elements and how they are represented in an engineering drawing.

9.1 KEYWAYS

Keys are machine elements which are used to prevent relative motion between two components of an assembly. As an example, this could be a key between a shaft and a hub to transmit torque. Types of different keys include the following:

- Square keys
- Flat keys
- Feather keys
- Taper keys
- Gib-head keys
- Woodruff keys

All these keys require a keyway, sometimes called a keyslot, in the shaft and the hub. The representation of such a keyway is exemplarily shown for a feather key. Note, keys are not internationally standardised, therefore refer to the national standards of your country. In this book the German standard DIN 6885–1 for feather keys is used.

The feather key is a parallel key with different possible variations in their form. One of the most commonly used feather keys is the type where both ends show a radius. The mating keyway in the shaft needs to be of the exact same shape and size; therefore, the keyway for a rounded feather key is machined with an end mill. There are two methods to adequately represent the keyway on a shaft in a drawing. In Figure 9.1 two different representations of a keyway

DOI: 10.1201/9781003001386-9

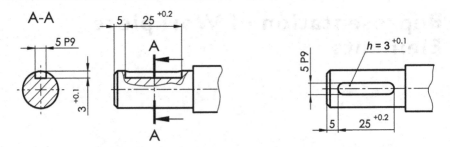

Figure 9.1 Two methods of how to represent and dimension a keyway.

in the end of a shaft are shown. On the left-hand side, the keyway is shown by using a broken-out section. The width and the depth of the slot is measured in a full section view. Alternatively, the slot can be shown in a regular view, but then the shaft must be rotated by 90° so that the plane of projection is parallel to the bottom face of the keyway (Figure 9.1, right-hand side). In this case, the depth of the keyway is dimensioned with the help of a leader line pointing to the bottom face. The value is preceded by $h =$ (which stands for 'height').

The advantage of the first method is that due to the separate section view there is more space for providing dimensions, tolerances, and maybe even surface texture requirements. However, it does not provide the true shape of the keyway. The second method includes just one view and, therefore, is more compact.

9.2 SPLINES AND SERRATIONS

Spline joints are used to coaxially connect two components in order to transmit a torque. This is an efficient shaft-hub joint which allows a certain axial movement and the transmission of high torque. Owing to its complex geometry, it is expensive to manufacture compared to alternative solutions. The shaft shows equally spaced teeth on its circumference. Accordingly, the hub has mating spaces. A typical spline joint can be seen in Figure 9.2.

Deriving the appropriate projection can be complex and time consuming. While modern 3D CAD systems automatically create the correct projections without problems (Figure 9.3), this can be hard work in manual drafting or even in a 2D CAD system.

For that reason, simplified representations are allowed to improve the efficiency of the drafting process. For example, ISO 6413 standardises the representation of splines and serrations, which is presented in the following.

Instead of drawing the complete representation, a simplified view on any projection plane passing through the shaft's axis eases the description of the spline joint. Figure 9.4 illustrates how a straight-sided spline is designated in a drawing. For clarity, the shaft is shown with a broken-out section to see how the spline is drawn in a section view.

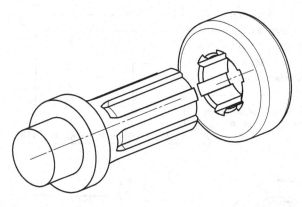

Figure 9.2 Spline joint.

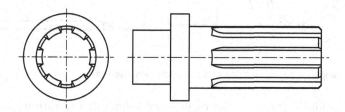

Figure 9.3 Full representation of a shaft with straight-sided spline.

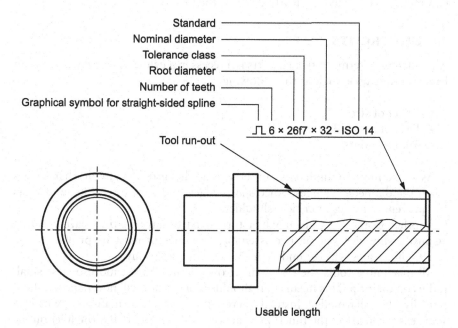

Standard
Nominal diameter
Tolerance class
Root diameter
Number of teeth
Graphical symbol for straight-sided spline

⊓ 6 × 26f7 × 32 - ISO 14

Tool run-out

Usable length

Figure 9.4 Simplified representation of a shaft with straight-sided spline.

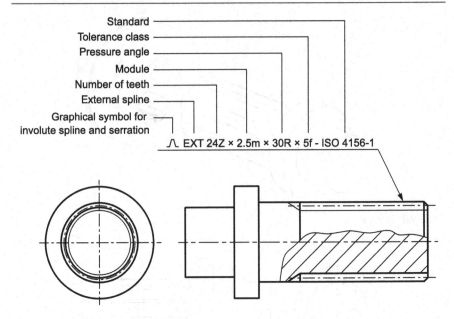

Figure 9.5 Simplified representation of a shaft with involute spline.

The simplified representation of involute splines and serrations looks very similar, except that in addition the pitch diameter is represented by a narrow dashed dotted line (Figure 9.5). The designation is also different because more parameters are needed to clearly describe the profile.

9.3 UNDERCUTS

An undercut in terms of turned parts is a recess in a cylindrical surface. This can have multiple uses, such as the following:

- Relief of stress
- Ease of assembly
- Tool run-out

As an example, an undercut can serve as a relief groove to avoid sharp edges when the diameter changes. Since sharp edges lead to stress peaks, undercuts can prevent damages at critical shoulders.

Another example is given when turned parts have mating parts, seating against a shoulder. For proper assembly, an undercut must be provided. Figure 9.6 illustrates why an undercut can facilitate assembly.

All machining tools such as a chisel on a lathe show a more or less small radius on their tip. This means that shoulders on a shaft will never show a theoretically ideal sharp edge. Instead, there will always be a small fillet, which in some cases can affect the function of an assembly. In Figure 9.6 the fillet on the shoulder prevents the hub from being flush with the lateral face of the shoulder.

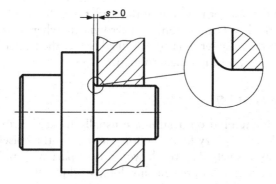

Figure 9.6 Shaft without undercut and a mounted hub.

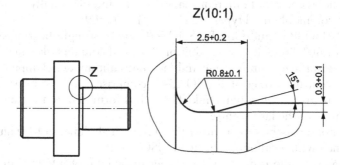

Figure 9.7 Full representation of an undercut.

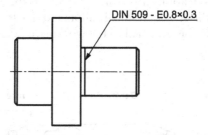

Figure 9.8 Simplified representation of an undercut.

Undercuts can be shown in full or as a simplified representation. The latter requires that the recess geometry be somehow standardised. Undercuts are not internationally standardised. As an example, this book uses the German standard for undercuts DIN 509. Undercuts exist for external geometry and for internal geometry such as holes as well. Without presenting the complete standard in detail, it suffices to point out that different types exist, depending on whether the two mating components show relative motion.

Figure 9.7 and Figure 9.8 depict an example of an external undercut shown in full representation, giving all dimensions in a detail view and in simplified form, giving the designation only.

Take care of the line termination of the leader line. Remember, when pointing to an edge the line termination is an arrowhead, whereas if the leader line points to a face then the terminator is a filled dot. If the leader line refers to a line (e.g. centre line), then no line termination is used.

9.4 CENTRE HOLES

Parts to be manufactured on a lathe are usually fixed in a three-jaw chuck. When the workpiece is very long, the force acting on the chisel would deflect the workpiece excessively. The result would be a part with reduced accuracy and poor surface finish. To prevent this, long workpieces are turned between centres, using a tailstock. To be able to turn a workpiece between centres, centre holes are required at the plane end faces.

Centre holes consist of different shapes and are internationally standardised. The most commonly used types are given in Figure 9.9.

Form A (ISO 866) and form B (ISO 2540) have a straight bearing surface at an angle of 60°. Form B has an additional conical countersink which protects the bearing surface from damage and facilitates automated clamping of parts due to the centring aid. Form R (ISO 2541) comprises a curved bearing surface without any protective countersink. The radius form allows a certain angular misalignment of the lathe centre.

Centre holes may be shown either in detail with all dimensions (Figure 9.10) or simplified with a callout according to ISO 6411.

The drawing callout depends on further machining. While in some cases it is accepted that the centre hole remains on the workpiece after the machining process, for some applications it is mandatory or maybe even undesirable. ISO 6411 suggests three different simplified callouts for centre holes which differ in the way the designation is attached to the workpiece (Figure 9.11).

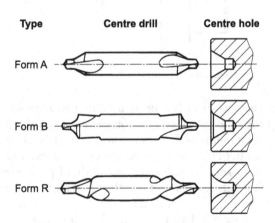

Figure 9.9 Centre drills and resulting centre holes for workpieces.

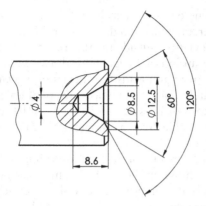

Figure 9.10 Shaft end with a centre hole form B shown in detail.

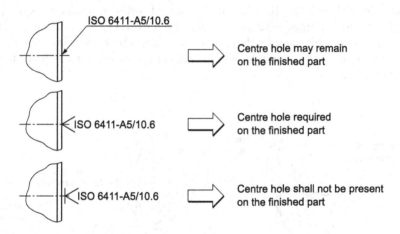

Figure 9.11 Simplified centre hole callouts according ISO 6411.

9.5 CORNERS AND EDGES

Technical drawings always show the theoretically ideal geometric shape. As stated in Chapter 7 and Chapter 8, after manufacturing the geometry differs from the ideal shape and shows more or less small imperfections. Dimensional and geometrical tolerances define the permissible variations of features in size, form, orientation, location and so on. So far, edges have not been considered and their accuracy is not controlled by any of the tolerances mentioned earlier. However, external edges can show burrs and internal edges a passing, caused by machining operations. These imperfections can lead to malfunctioning of a part or represent a safety issue.

ISO 13715 suggests how to control the edges of undefined shape by drawing indications.

Whenever deviations on geometrically ideal corners and edges occur, they can be either an undercut, which is a deviation inside the theoretical shape, or a passing, which is a deviation outside the theoretical shape. Figure 9.12 depicts an external edge and an internal edge, both drawn with an undercut, i.e. too much material is removed and leads to a deviation towards the material side. The thick dashed double-dotted line represents an alternative deviation, which is a passing. In this case there is too much material left and therefore it leads to a deviation to the outside. Such a passing is called a *burr* when emerging at an external edge.

To indicate the requirements for an edge according to ISO 13715, a graphical symbol is used as shown in Figure 9.13. This basic symbol can be used to explicitly specify external or internal edges only. To clearly indicate in a drawing, a symbol for an external or internal edge is added to the basic symbol.

To control whether an edge is allowed to show excess material or requires material removal, the basic symbol is supplemented by a symbol element. A plus sign (+) indicates that passing is allowed, while an undercut is not permitted. Whenever an edge requires material removal, i.e. an undercut, a minus sign (−) is used. This includes that any passing is not permitted. If either a passing or an undercut is permitted, a value indicating the edge size is preceded by the symbol ±.

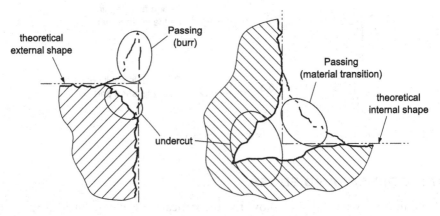

Figure 9.12 Definition of undercut and passing on theoretical edges.

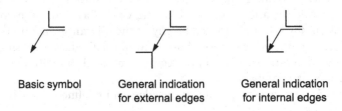

Figure 9.13 Basic graphical symbol for edge requirements.

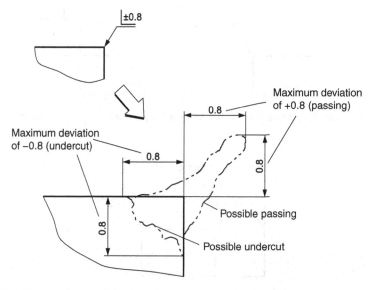

Figure 9.14 Definition of permissible edge sizes.

Edge sizes are located within the L-shaped symbol and indicate the maximum deviations in the principal directions. Figure 9.14 shows an example with an external edge. In this case the deviation can be positive or negative, i.e. a passing or an undercut is permitted. The maximum deviation is 0.8 mm each, as shown.

This convention applies similarly to internal edges, i.e. negative values for maximum undercut and positive values for maximum passing (burr).

Instead of indicating just one single value, similar to dimensional tolerances the deviation for an edge can be specified with an upper and a lower limit. If only one value is given, then this is considered to be the upper limit and the lower limit is 0. It is also possible to indicate permissible deviations for passing and for undercut in one symbol.

If a specific direction for the passing or undercut is required, the value for the edge size can be placed aligned with either the horizontal line or the vertical line. Note, specifying an edge size for one direction only is not possible for undercuts on external edges or passing on internal edges. These types require values for both directions.

The indication of direction is exemplarily shown for an external edge in Figure 9.15. If the passing on an external edge or an undercut on an internal edge should point in a vertical direction, then the edge size shall be aligned with the vertical line of the graphical symbol.

Edges in technical drawings can be individually indicated if they are perpendicular to the projection plane, i.e. in projection plane they appear as a corner.

If all edges of a given outline will be indicated, the graphical symbol should include a small circle at the kink of the leader line, indicating that the edge

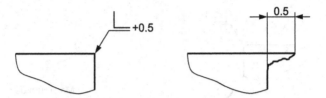

Figure 9.15 Indication of direction for the permissible passing on an external edge.

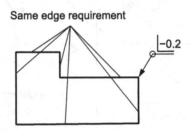

Figure 9.16 Indication of edge size for all edges around an outline.

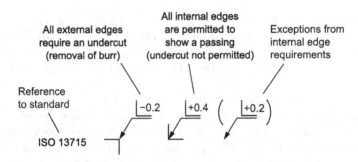

Figure 9.17 General indication of edge conditions near the title block.

requirement is for all edges the outline comprises. This convention is similar to the 'all around' symbol for surface texture requirements. In the example given in Figure 9.16, all marked edges are supposed to have an undercut of maximum 0.2 mm.

Symbols placed near the title block do not refer to specific edges but indicate that the requirement applies to all edges of the given type. Exceptions are to be indicated in parentheses. Figure 9.17 depicts an example.

Some examples of indication of undefined edges are given in Figure 9.18. The meaning is as follows:

(a) External edge without burr; material removal required between 0 and 0.2 mm; direction undefined

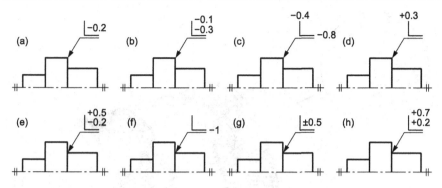

Figure 9.18 Drawing callouts for undefined edges.

(b) External edge without burr; material removal required between 0.1 and 0.3 mm; direction not specified

(c) External edge without burr; undercut required of size 0–0.4 mm in vertical direction and 0–0.8 mm in horizontal direction

(d) External edge with passing of size 0–0.3 mm in vertical direction permitted; undercut not permitted

(e) Internal edge; passing of size 0–0.5 mm permitted; undercut of size 0–0.2 mm permitted; direction not specified

(f) Internal edge with required undercut of up to 1 mm in horizontal direction

(g) Internal edge with permissible material removal between 0 and 0.5 mm, or permissible passing up to 0.5 mm; direction undefined

(h) Internal edge; passing of size 0.2–0.7 mm permitted; undercut not permitted; direction undefined

9.6 KNURLING

Whenever parts need enhanced grip, such as for knobs and adjustment screws, a special surface finish is applied called *knurling*. Grooves or ridged patterns are pressed into a round-shaped workpiece so that a firm grip is maintained. This helps to prevent a hand from slipping off during rotation. Knurling can also be applied for decorative reasons. Figure 9.19 illustrates a knurled machine element.

Different knurling patterns exist and need to be somehow indicated in an engineering drawing once required on a surface. As can be seen from Figure 9.19, representing a knurl in full detail can lead to excessive drafting work.

According to ISO 13444, knurling can be represented in simplified form. Figure 9.20 illustrates possible knurling patterns and how they are represented on an engineering drawing. Instead of filling the complete surface with the knurling pattern, it is only adumbrated and accompanied by a callout. The designation of the knurl according to ISO 13444 consists of the type of knurl

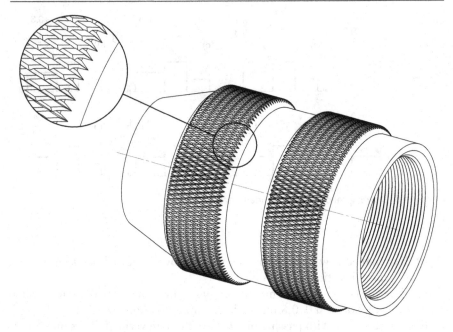

Figure 9.19 Nut with two knurled surfaces for improved grip.

which is only straight (type A) or diamond (type B). The reference letter for the type is followed by the diametral pitch of the knurling, which is the distance between the grooves.

In industry, further knurling patterns exist such as the right-/left-hand knurling and the cross knurling (Figure 9.20).

9.7 FURTHER CONVENTIONAL REPRESENTATIONS

Many common engineering details are difficult and tedious to draw. The screw thread is an example of this type of detail, and it will be shown in the subsequent chapter that there are conventional ways of drawing screw threads which are very much simpler than drawing out helical screw threads in full.

Figure 9.21 and Figure 9.22 show some more engineering details, and alongside the detailed drawing is shown the simplified representation for that detail. These conventions are designed to save time and should be used wherever and whenever possible.

These are not all the conventional representations that are available, but the rest are beyond the scope of this book. The interested student can find the rest in ISO 128 (in Great Britain, BS 308). There is a way of representing compression and tension springs diagrammatically. The coils of these springs can be represented by straight lines.

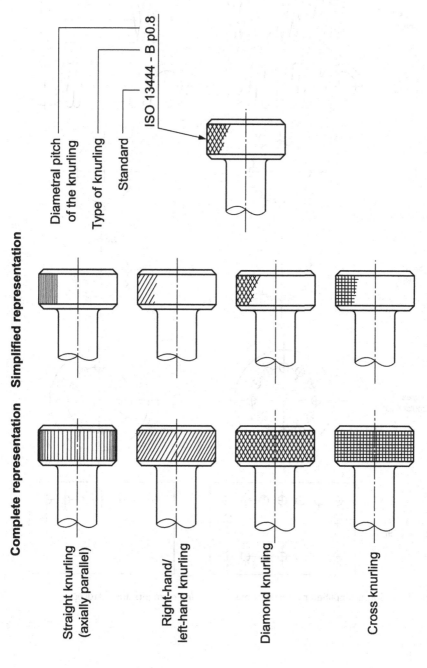

Figure 9.20 Representation forms for knurling patterns and their designation according to ISO 13444.

Title	Subject	Convention

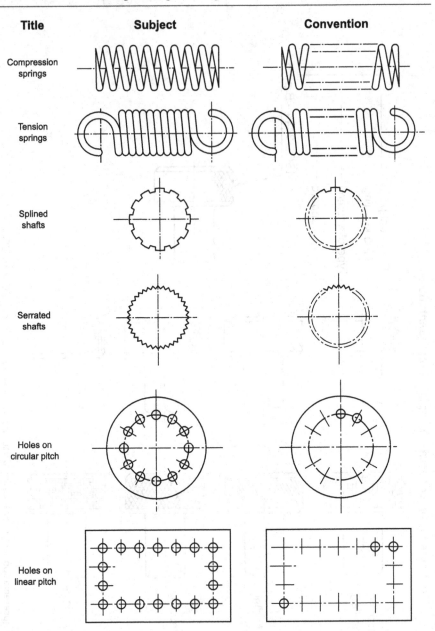

Figure 9.21 Some simplified representations of machine elements and features.

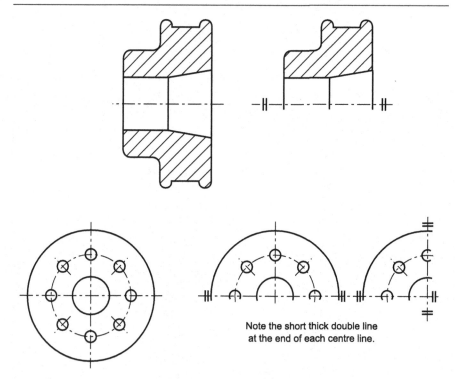

Note the short thick double line
at the end of each centre line.

Figure 9.22 Conventional representation of symmetrical parts.

Chapter 10

Representation of Machine Elements

10.1 THREADS

10.1.1 Screw Threads

The screw thread is probably the most important single component in engineering. The application of the screw thread to nuts, bolts, studs, screws, etc., provides us with the ability to join two or more pieces of material together securely, easily and, most importantly of all, not permanently. There are other methods of joining materials together, but the most widely used ones – riveting, welding and (very common these days) using adhesives – are all permanent. It is true that these methods are cheaper, but when we know that we might have to take the thing apart again, we use the screw thread. Since the screw thread is so important it is well worthwhile looking at the whole subject more closely.

The standard thread, in Great Britain and some other countries in Europe, for many years was the Whitworth; this thread was introduced by Sir Joseph Whitworth in the 1840s. It was the first standard thread; previously a nut and bolt were made together and would fit another nut or bolt only by coincidence. At the time, it was a revolutionary step forward.

The BSW (British Standard Whitworth) thread and its counterpart the BSF (British Standard Fine) thread were the standard threads in Great Britain until metrication and will probably be in use for many years.

However, the United States of America developed and adopted the unified thread as their standard and countries using the metric system of measurement had developed their own metric thread forms, such as the German metric DIN screw in 1919. It became increasingly obvious that an international screw thread was needed.

As far as the birth country of the modern screw was concerned, the breakthrough came when it was decided that British industry should adopt the metric system of weights and measures. The International Organization for Standardization (ISO) has formulated a complex set of standards to cover the whole range of engineering components.

Their thread, the ISO, is now the international standard thread. The ISO and unified thread profiles are identical. The unified thread is the standard international thread for countries which are still using imperial units.

DOI: 10.1201/9781003001386-10

The ISO basic thread form is shown in Figure 10.1.

You will note that the thread is thicker at the root than at the crest. This is because the stresses on the thread are greater at the root and the thread needs to be thicker there if it is to be stronger.

In practice, since there is nothing gained by having the root and crest of a nut and bolt in contact, and because 'square' corners are difficult to manufacture, the ISO thread form is usually modified to that shown in Figure 10.2.

You can see that the contact will be only on the flanks.

There is more than one type of ISO thread. There are 12 series of threads ranging from the widely used coarse thread series, which is used on bolts from 1.6 to 68 mm diameter, to a 6 mm constant pitch series with sizes from 70 to

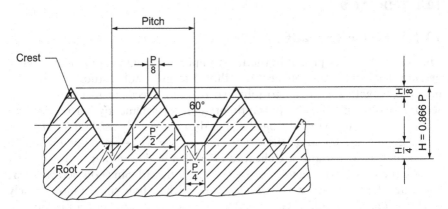

Figure 10.1 Basic form of ISO thread.

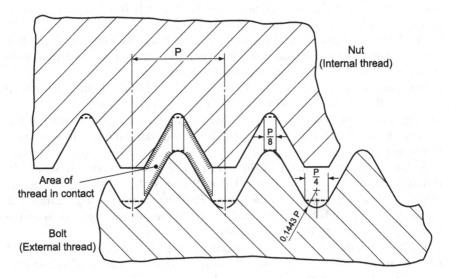

Figure 10.2 ISO thread used in practice.

300 mm diameter. The whole range of thread series has the same basic profile, and full details can be found in ISO 68–1. The fine thread series (the equivalent to the redundant BSF) ranges from 1.8 to 68 mm.

The British Standard Whitworth thread has now been superseded by the ISO thread. However, the ISO has adopted the Whitworth profile for pipe threads. It is called the British Standard Pipe Thread, Figure 10.3.

Some special threads have been designed to fulfil functions for which a vee thread would be inadequate. Some are shown.

The square thread, Figure 10.4, is now rarely used because it has been superseded by the acme thread. Its main application is for transmitting power since there is less friction than with a vee thread.

The acme thread, Figure 10.5, is extensively used for transmitting power. The thread form is easier to cut than the square thread because of its taper and, for the same reason, it is used on the lead screw of lathes where the half-nut engages easily on the tapered teeth.

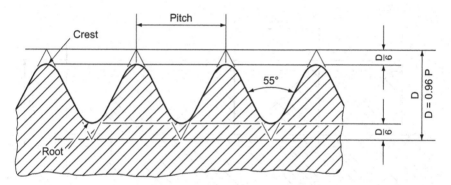

Figure 10.3 British standard pipe thread (British Standard Whitworth thread profile).

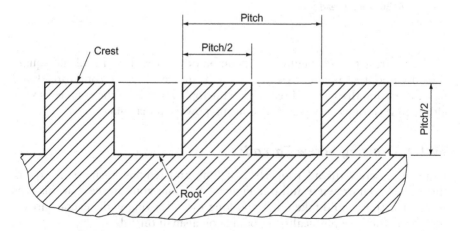

Figure 10.4 Square thread profile.

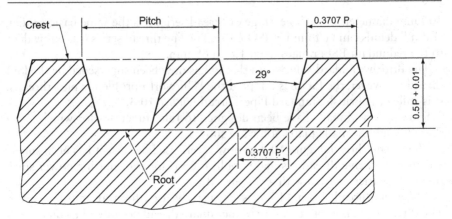

Figure 10.5 Acme thread profile.

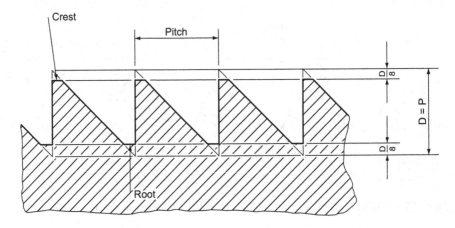

Figure 10.6 Buttress thread profile.

The buttress thread, Figure 10.6, combines the vee thread and the square thread without retaining any of their disadvantages. It is a strong thread and has less friction than a vee thread. Its main application is on the engineer's vice, although it is sometimes seen transmitting power on machines.

10.1.2 Drawing Screw Threads

Drawing a screw thread properly is a long and tedious business. A square thread has been drawn in full (see Figure 3.84), and you can see that this type of construction would take much too long for a drawing with several threads on it and would be physically impossible on a small thread.

There are conventions for drawing threads which make life very much easier. Three conventional methods of representing screw threads are shown in Figure 10.7. The top two methods are not used on engineering drawings any more. One of these illustrates the shape of the thread and the other has lines representing the thread crest and root. The bottom drawing shows how to draw a screw thread on an engineering drawing; the parallel lines represent the thread crest and root.

The only convention which shows whether the thread is right- or left-handed is the second one. This is not much of an advantage because the thread has to be dimensioned, and it is a simple matter to state whether a thread is right- or left-handed. Left-hand threads are rarely met with and, unless specifically stated, a thread is assumed to be right-handed.

Figure 10.7 shows the convention for both external and internal threads. It should be explained that, on the drawings for internal threads, the thread does not reach to the bottom of the hole. When an internal thread is cut, the material is first drilled a little deeper than is actually required. The diameter of the hole is the same as the root diameter of the thread and is called the tapping diameter. The screw thread is then cut with a tap, but the tap cannot reach right to the bottom of the hole and some of the tapping hole is left. The cutting angle of the drill, for normal purposes, is 118° to almost 120°. Thus, a 60° set square is used to draw the interior end of an internal screw thread.

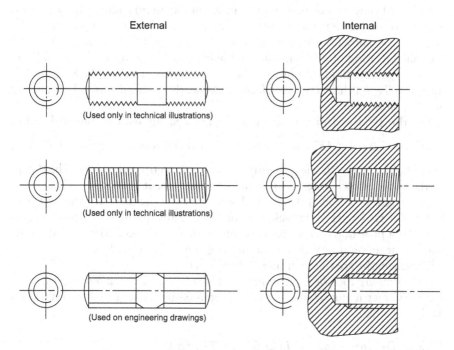

External   Internal

(Used only in technical illustrations)

(Used only in technical illustrations)

(Used on engineering drawings)

Figure 10.7 Conventional representation of screw threads.

10.2 FASTENERS

10.2.1 Screws, Bolts and Studs

The most widespread application of the screw thread is the nut and bolt show-ing full details to whenever a nut or bolt is drawn; it is essential that the first view drawn is the one which shows the regular hexagon. If the across-flats (A/F) dimension is given, draw a circle with that diameter. If it is not, look in an engineering handbook, a book of tables or the corresponding standard under the appropriate thread size. Construct a regular hexagon round the circle with a 60° set square. Project the corners of the hexagon onto the side view of the nut and bolt and mark off the thickness of the nut or bolt head.

Nuts and bolts are chamfered and, when viewed from the side, this chamfer is seen as radii on the sides of the nut or bolt. If you ensure that the first view projected from the hexagon is the one which shows three faces of the nut (the other view shows only two faces), you can draw a radius equal to D, the diam-eter of the thread, on the centre flat. The intersection of this radius and the cor-ners of the neighbouring flats determines the size of the two smaller radii. These must start at this intersection, finish at the same height on the next corner and touch the top of the nut or bolt at the centre of the flat. This may be done by trial and error with compasses or with radius curves. Remember that the centre of the radius lies midway between the sides. This view is completed by drawing the 30° chamfer which produced the radii.

The third view of the nut or bolt is drawn in a similar fashion. The width and heights are projected from the two existing views, and the radii are found in the same way as shown on the other view. However, ISO 6410–3 recommends a simplified method of drawing nuts and bolts, and this is shown in Appendix B.

Two types of standard nuts and bolts are shown in Figure 10.8. Type A is standard. Type B has a 'washer face' underneath the head of the bolt and on one face of the nut.

The threaded end of the bolt may be finished off with a spherical radius equal to $1\frac{1}{4}D$ or a 45° chamfer to just below the root of the thread. Both of these enable the nut to engage easily and leave no sharp projections. The thread on the nut is also chamfered to assist easy engagement.

The length of a bolt is determined simply by the use to which the bolt is to be put. There is a very large selection of bolt lengths for all diameters. The bolt should not protrude very far past the nut, so there is no need to thread all of the shank. The amount of thread on a bolt is given in Table 10.1.

These lengths are the minimum thread lengths.

Bolts that are too short for minimum thread lengths are called screws.

An ISO metric nut or bolt is easily recognised by the letter 'M' or ISO M on the head or flats.

10.2.2 Designation of ISO Screw Threads

The coarse series ISO thread is only one of 12 different threads in the ISO series. This thread, like the fine thread series, has a pitch which varies with the

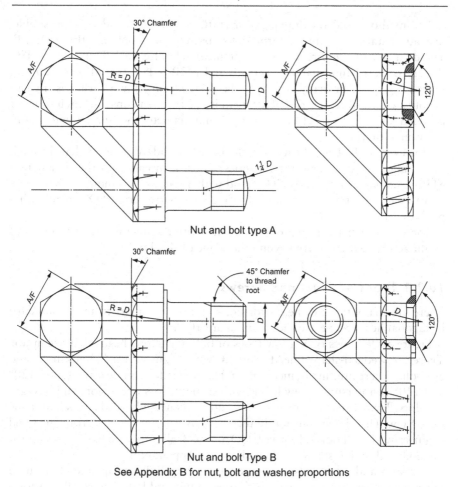

Nut and bolt type A

30° Chamfer

45° Chamfer
to thread
root

Nut and bolt Type B
See Appendix B for nut, bolt and washer proportions

Figure 10.8 Two types of standard nuts and bolts.

Table 10.1 Amount of thread on a bolt

Length of bolt	Length of thread
Up to and including 125 mm	2d + 6 mm
Over 125 mm and up to and including 200 mm	2d + 12 mm
Over 200 mm	2d + 25 mm

d is the diameter of the bolt.

diameter of the bolt. The remaining ten thread series have constant pitches, whatever the diameter of the thread.

All the series except the coarse thread series are used in special circumstances. The vast majority of threads used come from the coarse thread series.

The method used on drawings for stating an ISO thread is quite simple. Instead of stating the thread form and series, you need only use the letter 'M'. The diameter of the thread is stated immediately after 'M'. Thus M12 is ISO thread form, 12 mm diameter thread, and M20 is ISO thread form, 20 mm diameter thread. In many countries the designation shown earlier is used to denote a coarse thread series. If a thread is used from a constant pitch series, it is added after, so that M14 × 1.5 is a 14 mm diameter ISO thread with a constant pitch of 1.5 mm.

However, in the United Kingdom the British Standard requires that the pitch be included in the coarse thread series. Thus, a thread with the designation M30 × 3.5 is a coarse series ISO thread with a pitch of 3.5 mm.

A thread with the designation M16 × 2 is a coarse series ISO thread with a pitch of 2 mm.

There are further designations concerned with the tolerances, or accuracy of manufacture, but these are beyond the scope of this book.

10.2.3 Types of Bolts and Screws

There are many types of heads for bolts and screws apart from the standard hexagonal head. Some are shown in Figure 10.9.

Figure 10.9 shows only a few types of bolt and screw heads that are in use. There are wedged-shaped heads, tommy heads, conical heads, hook bolts and eye bolts. There are small-, medium- and large-headed square screws, 60°, 120° and 140° countersunk screw heads with straight slots, cross slots and hexagonal slots. There are instrument screws and oval cheese-headed screws, to name only a few. The dimensions for all these screws can be obtained from any good engineering handbook. However, ISO 6410–3 also recommends a simplified method of drawing screws. This is shown in Appendix B.

The stud and set bolt (sometimes called a tap bolt or cap screw) are used when it is impossible or impractical to use a nut and bolt. Figure 10.10 shows both in their final positions. They are both screwed into a tapped hole in the bottom piece of material. The top piece of material is drilled slightly larger than the stud or screw and is held in position by a nut and washer in the case of the stud, and by the head of the set bolt and washer in the case of the set bolt. The stud would be used when the two pieces of material are to be taken apart quite frequently; the set bolt would be used if the fixing is expected to be more permanent.

10.2.4 Locking Devices

Constant vibration tends to loosen nuts and, if a nut is expected to be subjected to vibration, a locking device is often employed. There are two basic groups of locking devices: one group increases the friction between the nut and the bolt or stud; the other group is more positive and is used when heavy vibration is anticipated or where the loss of a nut would be catastrophic. Figure 10.11 shows five common locking devices.

First angle projection

Bolt heads

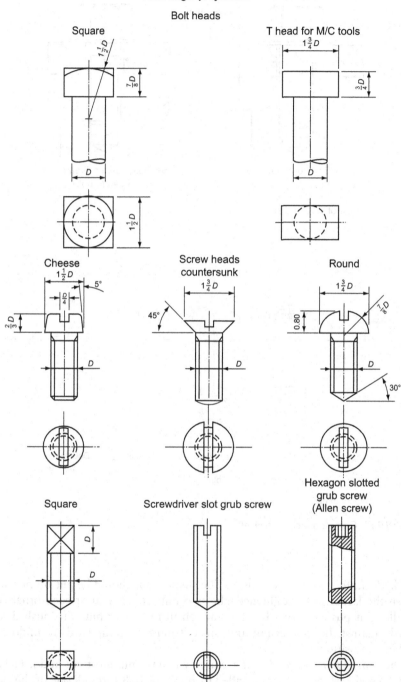

Figure 10.9 Bolt heads and screw heads.

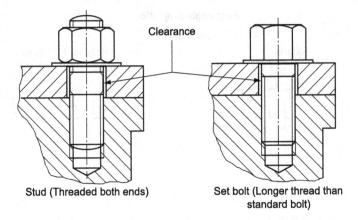

Stud (Threaded both ends)   Set bolt (Longer thread than
                                             standard bolt)

Figure 10.10 Use of stud and set bolt.

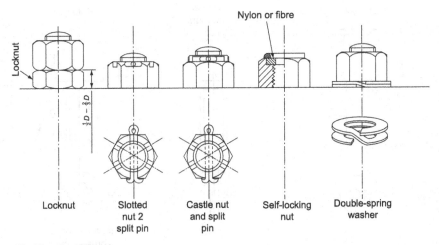

D = Diameter of thread

Figure 10.11 Common locking devices.

The *locknut* is very widely used. The smaller nut should be put on first and, when the larger nut is tightened, the two nuts strain against each other. The smaller nut pushes upwards. The reaction in the larger nut is to push downwards against the smaller nut and, since it must move upwards to undo, it is locked in position.

The *slotted nut* and *castle nut* are used when the nut must not undo. The nut is tightened and then a hole is drilled through the bolt through one of the slots. A split pin is inserted and the ends are bent over. A new split pin should be used each time the nut is removed.

The *self-locking* nut is now very widely used. The nylon or fibre washer is compressed against the bolt thread when the nut is tightened. This nut should only be used once since the nylon or fibre is permanently distorted once used.

The *spring washer* pushes the nut up against the bolt thread, thus increasing the frictional forces. It is the least effective of the locking devices shown and should only be used where small vibrations are expected.

There are many other types of locking device and full descriptions can be found in any good engineering handbook.

10.2.5 Rivets and Riveted Joints

A rivet is used to join two or more pieces of material together permanently. The enormous advances in welding and brazing techniques and the rapidly increasing use of bonding materials have led to a slight decline in the use of rivets. However, they remain an effective method of joining materials together, and, unlike welding and bonding, require very little special equipment or expensive tools when used on a small scale.

The rivet is usually supplied with one end formed to one of the shapes shown in Figure 10.12. The other end is hammered over and shaped with a tool called a 'dolly'.

When rivets are used, they must be arranged in patterns. The materials to be joined must have holes drilled in them to take the rivets; these holes weaken the material, particularly if they are too close together. If the rivets are placed too close to the edge of the material, the joint will be weakened. The two basic joints are called 'lap and butt joints'. Figure 10.13 shows four examples. There is no limit to the number of rows of rivets or to the number in each row, but the spacing, or pitch of the rivets, must be as shown.

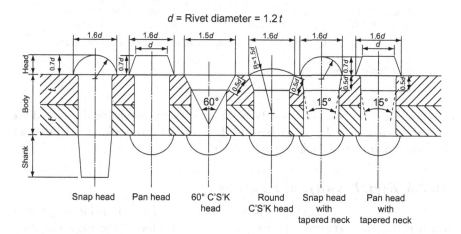

Figure 10.12 Types of rivet forms.

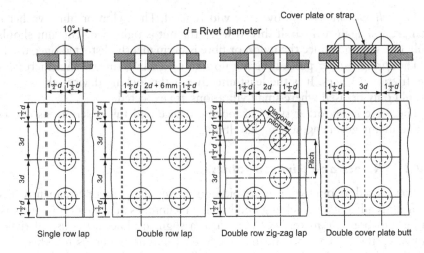

Figure 10.13 Arrangement of rivets in patterns.

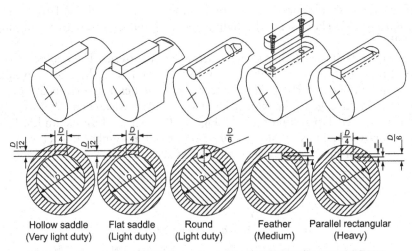

| Hollow saddle | Flat saddle | Round | Feather | Parallel rectangular |
| (Very light duty) | (Light duty) | (Light duty) | (Medium) | (Heavy) |

Proportions are approximated for drawing only. For exact dimensions see ISO TC/16.

Figure 10.14 Types of keys.

There are other types of rivet, the most important group being those used for thin sheet materials. These are beyond the scope of this book, and details can be found in any good engineering handbook.

10.2.6 Keys, Keyways and Splines

A key is a piece of metal inserted between the joint of a shaft or hub to prevent relative rotation between the shaft and the hub. One of the most common applications is between shafts and pulleys.

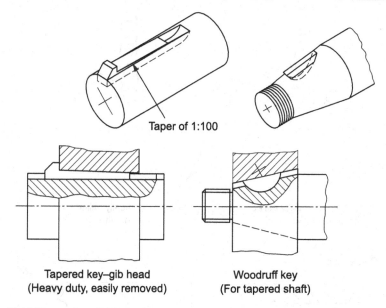

Taper of 1:100

Tapered key–gib head
(Heavy duty, easily removed)

Woodruff key
(For tapered shaft)

Figure 10.15 Tapered key – gib head (heavy duty, easily removed) and Woodruff key (for tapered shaft).

There is a wide variety of keys, designed for light and heavy duties, for tapered and parallel shafts and to allow or prevent movement of the hub along the shaft (Figure 10.14 and Figure 10.15).

Saddle keys are suitable for light duty only, since they rely on friction alone.

Round keys are easy to install because the shaft and hub can be drilled together, but they are suitable for light duty only.

Feather keys and *parallel keys* are used when it is desired that the hub should slide along the shaft, yet not be allowed to rotate around the shaft.

Taper keys are used to prevent sliding, and the *gib head* allows the key to be extracted easily.

Woodruff keys are used on tapered shafts. They adjust easily to the taper when assembling the shaft and hub.

Keyways are machined out with milling machines. If a horizontal milling machine is used, the resulting keyway will look like the one to the left in Figure 10.16. If a vertical milling machine is used, the resulting keyway will look like the one to the right in Figure 10.16. In both cases, the end of the milled slot has the same profile as the cutter.

If a shaft is carrying very heavy loads, it should be obvious that the load is transferred to the hub (or vice versa) via the key. This means that the power that any shaft or hub can transmit is limited by the strength of the key. If heavy loading is expected, the shaft and hub will be splined (Figure 10.17). The number of splines will be dependent upon the load to be carried; the greater the number of splines, the greater the permissible loading.

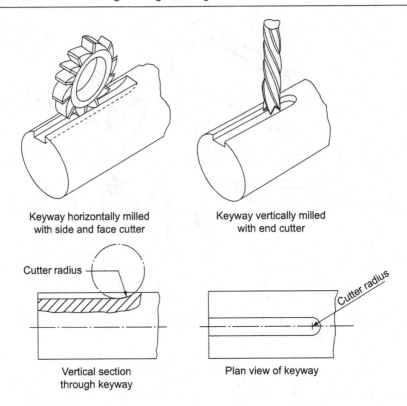

Keyway horizontally milled
with side and face cutter

Keyway vertically milled
with end cutter

Cutter radius

Vertical section
through keyway

Plan view of keyway

Cutter radius

Figure 10.16 Machining of a keyway.

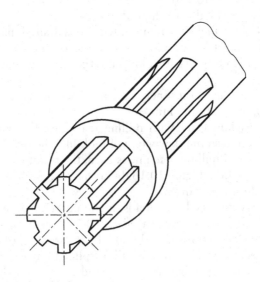

Figure 10.17 Splined shaft and hub.

10.2.7 Cottered Joints

Keys and splines are used when shafts are subjected to torsional (twisting) loads. If two shafts have to be joined together and then subjected to tension or compression (push–pull), a different type of fastening is needed. One method is to use a cotter. Two examples of cottered joints are shown in Figure 10.18. On the left is a connection for two round shafts, and on the right is the connection for two square or rectangular shafts.

The whole assembly locks together as follows.

As the tapered cotter is forced downwards, it reacts against faces A and B and tries to draw the shafts together. The smaller shaft cannot be pulled in any further, either because of the collar at C or because the shafts meet at D. Thus, the more the cotter is forced down, the tighter the assembly. The shafts are easily separated by knocking out the cotter from underneath.

The square or rectangular bar is opened out to form a Y-shaped fork. The smaller bar fits inside the arms of this Y. A gib is used to prevent the arms from spreading when the cotter is hammered in. This problem does not arise on round bars because the larger bar wraps completely round the smaller one.

10.3 PROBLEMS

(All questions originally set in imperial units.)

1   Figure 10.19 shows two views, in first angle projection, of the body and base of a shaft bearing. The body is secured in position on the base by two hexagon-headed M8 set screws, 25 mm long, complete with washers. Draw, full size and in correct projection, the following views of the complete assembly, together with a length of 25 mm diameter shaft in

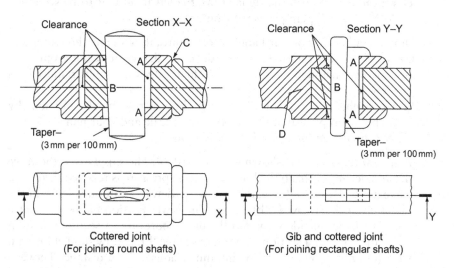

Figure 10.18 Examples of cottered joints.

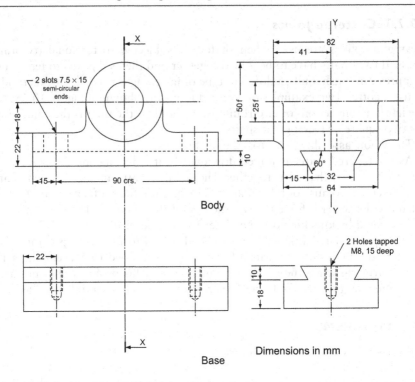

Figure 10.19 Body and base of a shaft bearing.

position, approximately 150 mm long and shown broken at both ends: (a) a sectional side elevation on X–X in the direction indicated; (b) an elevation projected to the right of (a), the left-hand half to be sectioned on Y–Y; (c) a full plan projected from (a).

Draw in either first or third angle projection stating the method adopted. Use your own discretion where any detail or dimension is not given.

Do not dimension your drawing; hidden edges are only required on view (c).

In a convenient corner of your paper draw a title block, 115 mm by 65 mm, and print in it the drawing title, scale and your name.

University of London School Examinations

2 A special relief valve is shown in Figure 10.20. The supply is at the valve 'D' and delivery is from the bore 'B'. Excess pressure is relieved through a valve at 'R'. Draw, full size, the following views: (a) an outside FE in the direction of the arrow *F*; (b) a plan. The lower portion below the centre line 'CC' is an outside view and the upper portion is a section taken on the plane 'PP'; (c) a sectional EE on the plane 'EE'. Show hidden lines on view (c) only. Insert the following dimensions: (i) the distance between

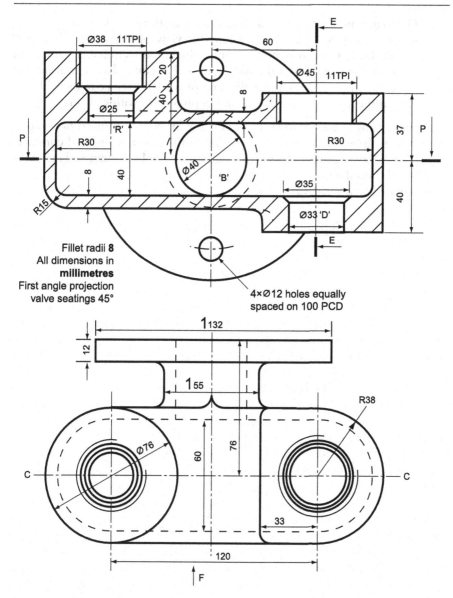

Figure 10.20 Special relief valve.

each valve and the delivery bore; (ii) the diameter of the flange; (iii) the distance between the face of the flange and the centre line of the valves.

Complete in a suitable title block, along the lower border of the paper, the title, scale, system of projection used and your name.

Oxford and Cambridge Schools Examination Board

3 Orthographic views of a casting, drawn in third angle projection, are shown in Figure 10.21. Do *not* copy the views as shown but draw, full size in third angle projection, the following: (a) a sectional elevation, the plane of the section and the direction of the required view indicated at X–X; (b) an EE as seen in the direction of arrow E; (c) a complete plan as seen in the direction of arrow P and in projection with view (a).

Hidden edges are *not* to be shown on any of the views. Insert *two* important dimensions on each view, and in the lower right-hand corner of the drawing paper draw a title block 115 mm by 65 mm and insert the relevant data.

Cambridge Local Examinations

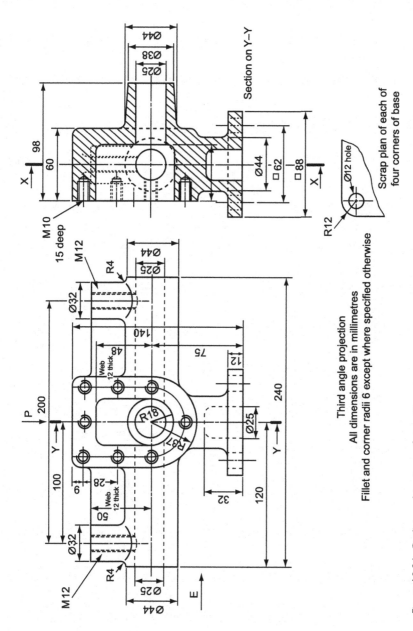

Figure 10.21 Orthographic views of a casting.

Chapter 11

Assembly Drawings

Few engineering items are completely functional on their own. There are some, a spanner or a rule for instance, but even a simple object like a wood chisel has three components, and a good pair of compasses may have 12 component parts. Each part should be drawn and dimensioned separately, and then a drawing is made of all the component parts put together. This is called an 'assembly drawing'.

11.1 INTRODUCTION

Engineering drawings play a crucial role in product development since they help us communicate, visualise and document a design. However, drawings alone will not fully describe a product. Nowadays, 3D CAD models, reports, calculation sheets, change orders, etc., are part of the product documentation. Nevertheless, a working drawing is one of the most important documents since it fully describes a product's shape and size and specifies the manufacture and assembly of the components.

A set of working drawings usually comprises detail drawings, i.e. multi-view drawings of single parts which specify the part's geometry, material, surface finish and sometimes even manufacturing steps. In case of assemblies, one or more assembly drawings are additionally required to show the relative position and orientation of the parts and how they interact.

The number and type of assembly drawings depend on the complexity of the product and therefore on its product structure. An assembly can consist of components which are grouped together to a subassembly but also of single parts. In Figure 11.1, a product structure is given. On the uppermost level the product consists of two assemblies and two single parts. The assemblies on their part comprise single parts and subassemblies. Single parts can be assembled multiple times and therefore occur in different levels of the product structure such as Part 4.

Each component, which is not a standard part that is purchased from a supplier, requires a detail drawing for manufacturing.

Assemblies and subassemblies are presented in assembly drawings and usually consist of the parts shown in their operating position, including item

DOI: 10.1201/9781003001386-11

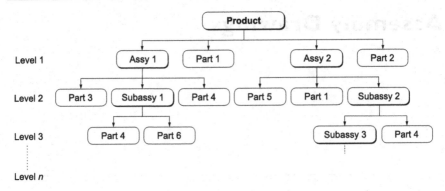

Figure 11.1 Example of a product structure.

numbers for identification and a parts list with some basic information. Item numbers and parts lists are thoroughly discussed in the following.

Figure 11.2 shows a typical assembly drawing.

Apart from the presented general assembly drawing which shows an overall assembly, there are further types of assembly drawing:

- *Outline assembly drawing*, showing the exterior envelope and overall dimensions only
- *Subassembly drawing*, showing just a group of components and allowing for fewer details to be shown in the overall assembly drawing
- *Exploded assembly drawing*, depicting the components separated by distance along their principal mounting direction to show the relationship between them or the sequence of assembly
- *Pictorial assembly drawing*, showing all parts separated in a pictorial view to illustrate how they are assembled (used in maintenance manuals, for example)
- *Installation drawing*, showing all components in their correct operating position and providing information on how to erect or assemble a machine
- *Fabrication drawing*, depicting the relative position of components of an assembly whose constituents are permanently joined, e.g. welding drawing
- *Schematic assembly drawing*, using standardised symbols and conventional representations only to illustrate the assembly
- *Layout drawing*, showing the final position of assemblies or components relative to the installation site, i.e. indicating buildings, structures, already existing machine components, etc.
- *Working* or *detailed assembly drawing*, fulfilling the function of detail drawings and an overall assembly drawing due to combination of the provided information (usually not recommended)

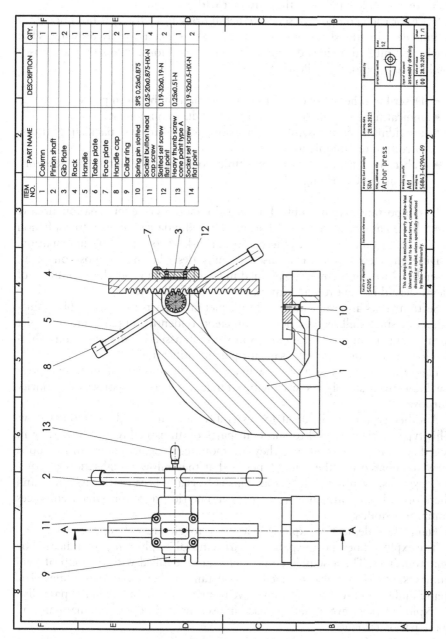

ITEM NO.	PART NAME	DESCRIPTION	QTY.
1	Column		1
2	Pinion shaft		1
3	Gib Plate		2
4	Rack		1
5	Handle		1
6	Table plate		1
7	Face plate		1
8	Handle cap		2
9	Collar ring		1
10	Spring pin slotted	SPS 0.25x0.875	1
11	Socket button head cap screw	0.25-20x0.875-HX-N	4
12	Slotted set screw flat point	0.19-32x0.19-N	2
13	Heavy thumb screw cone point type A	0.25x0.51-N	1
14	Socket set screw flat point	0.19-32x0.5-HX-N	2

Figure 11.2 Assembly drawing of an arbor press.

11.2 ASSEMBLY VIEWS

Assembly drawings are used to describe the relationship between parts, i.e. how they are assembled and how they interact with each other.

Since for each component separate detail drawings exist, it is not important to fully describe the shape and size of each part in an assembly drawing. That is why assembly drawings do not show any dimensions, unless they are absolutely necessary for the following:

- Describing the relative position and orientation between parts
- Indicating the range of motion for individual components
- Providing overall assembly dimensions needed for e.g. packaging, transportation or commissioning
- Fabrication instructions or machining operations required for assembly, such as match drilling

It is important that assembly drawings do not duplicate information already provided on any other drawing. This kind of redundancy can lead to confusion and, in case of modifications, there is even a risk of contradictory information.

Different view types can be used to fully describe the composition of an assembly. Most of them are explained in Chapter 6. In general, the number of views should be kept to a minimum to increase legibility.

Section views are frequently used to depict inner parts of an assembly. Sometimes one single full section view is sufficient to identify all parts and to depict their arrangement. When using section views in assembly drawings, omit hidden lines unless they are necessary for part identification.

Figure 11.3 shows a section view in an assembly drawing of an air compressor. Note, the assembly is shown without item numbers and parts list for clarity purposes.

Another type of view, ideal for assembly drawings, is the exploded view. This type of view is used to show all parts of an assembly and how they fit together or in which sequence they are mounted. The advantage of exploded views compared to other view types is that they allow visualisation of complex systems, even for a layperson. This is basically because the assembly and therefore all parts are shown as an axonometric projection which enhances comprehensibility.

Figure 11.4 depicts an exploded view of a toggle press clamp.

The explode lines, i.e. the paths of part removal (explosion), are indicated by phantom lines. The assembly components are identified with the help of item numbers, sometimes also referred to as detail numbers, and leader lines. The item numbers in the drawing view have to match the numbers in the parts list.

Exploded views are frequently used in user manuals, operation manuals and spare part catalogues.

Apart from full section views and exploded views, assembly drawings can consist of all types of views which also can be used for detail drawings. The

A-A   A

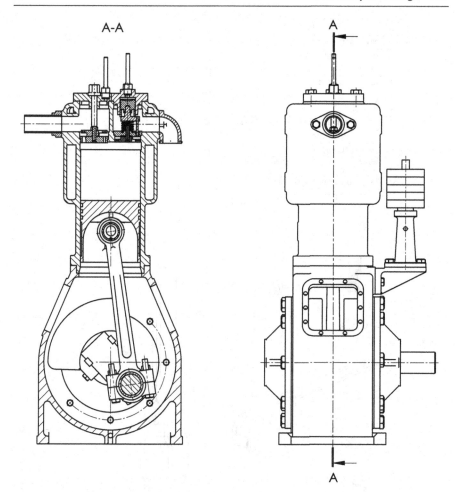

Figure 11.3 Full section view of an air compressor.

same rules for conventional representation, dimensioning, tolerancing and annotating apply.

11.3 PARTS LISTS

Assembly drawings are an important part of product documentation. However, the drawing alone is not sufficient to fully describe the product structure, to identify standard parts, or to figure out where to get the detail drawing of individual components from. Another important piece of information is the *parts list* or *bill of materials (BOM)*.

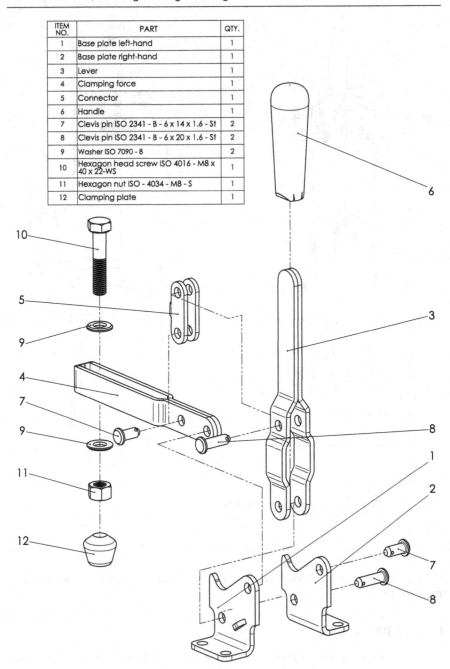

ITEM NO.	PART	QTY.
1	Base plate left-hand	1
2	Base plate right-hand	1
3	Lever	1
4	Clamping force	1
5	Connector	1
6	Handle	1
7	Clevis pin ISO 2341 - B - 6 x 14 x 1.6 - St	2
8	Clevis pin ISO 2341 - B - 6 x 20 x 1.6 - St	2
9	Washer ISO 7090 - 8	2
10	Hexagon head screw ISO 4016 - M8 x 40 x 22-WS	1
11	Hexagon nut ISO - 4034 - M8 - S	1
12	Clamping plate	1

Figure 11.4 Exploded view of a toggle press clamp.

Parts lists are usually given in table form and provide some basic information about the individual components of an assembly. Standardised parts lists (e.g. according to ISO 7573 or ASME Y14.34) make sure that the layout and the structure from different assembly drawings are similar, no matter in which company or country it is used. However, small differences such as the number of columns and their sequence can occur. As a rule, a parts list comprises the following data fields:

- A 'Part reference' or 'Identifier' enables to link the constituents of an assembly given in the drawing with the items in the parts list. The part reference is sometimes called *item number* or *detail number*.
- The data field 'Quantity' describes how many pieces of the respective part are used in the assembly. Quantity can also refer to the amount of material given in e.g. volume or length units if quantification in pieces is not applicable.
- The field 'Unit' expresses the standard of measurement. Usually parts are given in pieces. However, constituents can also be given in other units such as kg, m and l.
- A 'Part name', which is the designation of the part.
- 'Description' provides any technical designation used for manufacturing, suppliers, or identification of standard parts. In some parts lists this field is combined with the field 'Part name'.
- 'Part number' or 'Drawing number' serves as a unique identifier within a company.

Depending on a company's internal organisation, parts lists can further include data fields related to the material of the components, the weight which usually is important for transportation, and additional remarks to further describe a component.

Parts lists can represent a separate document or be included in an assembly drawing. Parts lists consisting of a few components only can be easily placed on the drawing sheet, either in conjunction with the title block or at any position on the drawing sheet. When the parts list is in conjunction with the title block, the table header shall be located at the bottom of the list (see Figure 11.5). If the parts list is placed elsewhere or is a separate document, then the header can be given on top (compare Figure 11.2) or at the bottom of the table. Sticking to this convention allows a parts list to be expanded, i.e. adding further parts, without renumbering all components.

As a rule, a parts list is sorted by size (largest parts first) or by importance (most important parts first). As an example, the housing of a gearbox would be listed first, and standard parts such as screws, washers and nuts would be located at the end of the list.

11.4 ITEM NUMBERING

Each item number in the parts list is assigned to a component so that all parts in the drawing can be clearly identified. Components which are assembled

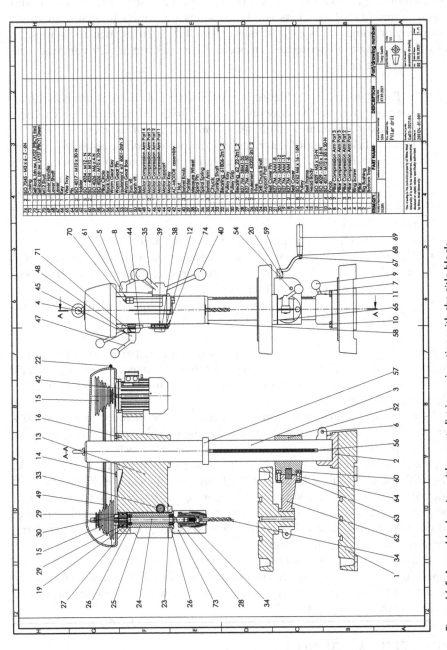

Figure 11.5 Assembly drawing with a parts list in conjunction with the title block.

multiple times have the same item number, unless they have different configurations. As an example, two hexagon head screws ISO 4014 – M10 × 60 – 8.8 will have the same item number, and in the parts list the quantity is set to 2. If, for example, the length would differ, then both screws would have different item numbers since the configuration would not be the same anymore.

To clearly differentiate item numbers from other annotations or callouts on a technical drawing, item numbers are given in letters twice the height of standard text (e.g. dimensions, annotations). As a rule this means a height of 7 mm.

An additional or alternative technique is to provide item numbers in balloons, i.e. in circles of continuous narrow lines.

Item numbers are placed outside any outline and point to the corresponding part with the help of leader lines. The leader line is usually drawn slanted. Horizontal or vertical orientation should be omitted to avoid any confusion. If necessary, a leader line can be drawn sharp bended.

Each leader line shall have a filled dot as line termination if it points to a face of a part. Sometimes the faces are too small to place a filled dot. In such case the leader line termination is a filled arrowhead and points to an edge. Whenever a leader line points to a line, e.g. a centre line of a standard part, no line termination is used. Figure 11.6 illustrates the rules previously mentioned.

The item numbers should be arranged in a legible manner. Usually the item numbers are horizontally aligned on top of the drawing view. However, for small assemblies the item numbers can be arranged clockwise around the outline. For large assemblies they can also be arranged in a rectangular.

If needed and if there is no risk of misinterpretation, item numbers can also be grouped. As an example, whenever two parts are joined with screws, usually washers and nuts are also part of the joint. In this case the item numbers for these parts are grouped and appear in their sequence of assembly, separated by a comma. Grouping of items can be helpful when the parts are considered to form a functional group. It eases identification and the search inside the parts list.

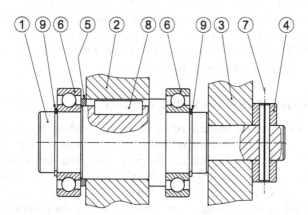

Figure 11.6 Assembly shown with leader lines and item numbers.

If identical parts occur multiple times in an assembly, it is sufficient to indicate the corresponding item number just once. There is no need to indicate the item number for each instance unless it increases legibility and avoids confusion. Especially for small assemblies the identification of identical parts without item number should be possible. For large assemblies it can ease reading the drawing if the item number for multiple instances is separately shown. Especially when the identical parts appear in different views.

11.5 REVISION BLOCK

In every product development process there is a chance of making mistakes. As long as they happen during the design process they can be corrected with limited effect on downstream processes. However, despite double-checking of drawings, mistakes can also become apparent only after drawings have already been released. In such a case the drawing needs to be corrected and the change is to be documented. Further reasons for documenting such changes on a drawing can be due to changing customer requirements or issues that lead to a new situation for the company. As an example, this can be a lack of availability of materials or third-party supplier components, or a change in the manufacturing equipment which requires a redesign. When designs exist for years, changes might be necessary to improve the function of a part or to reduce cost.

Changes, wherever they originate from, are usually documented in separate documents such as 'change notices' or 'change requests'. If the change impacts the design, the drawing is to be revised and all modifications are documented on the drawing in a 'revision block'.

In a revision block changes are listed in table form. For each change it is advisable to indicate a revision number, the date of change, the person who did the changes on the drawing, a short description of what exactly has changed in the drawing, the approver's name, and a reference to the change notice, if applicable.

Revisions are identified by consecutive numbers, each number for one change process. Since a revision reflects a status after modifications have been done, a unique identifier of a drawing always includes the revision number.

As soon as a new revision is out, the drawing with the previous revision number should be locked for further use and archived to avoid parallel circulation and therefore contradictory information.

11.6 READING ASSEMBLY DRAWINGS

Students at school or college are often instructed to draw the assembled components only and are shown the dimensioned details in no particular order. If the assembly is particularly difficult, the parts are often shown in an exploded view and the assembly presents no difficulty. The assembled parts may form an object which is easily recognisable, but the real problem occurs when there seems to be no possible connection between any of the component parts. In an examination, when loss of time must be avoided at all costs, the order of assembly needs to be worked out quickly.

The only approach is to view the assembly somewhat like a jigsaw puzzle. The parts must fit together and be held together, either because they interlock or there is something holding them together.

Learn to look for similar details on separate components. If there is an internal square thread on one component and an external thread of the same diameter on another component, the odds are that one screws inside the other. If two different components have two or more holes with the same pitch, it is likely that they are joined at those two holes. A screw with an M10 thread must fit an M10 threaded hole. A tapered component must fit another tapered component.

The important thing, particularly in examinations, is to start drawing. Never spend too long trying to puzzle out an assembly. There is always an obvious component to start drawing, and, while you are drawing that, the rest of the assembly will become apparent as you become more familiar with the details.

11.7 PROBLEMS

(All questions originally set in imperial units.)

1. Figure 11.7 shows a detail from a stationary engine. Draw this detail with the parts assembled. You may use either first or third angle projection.

   Draw, twice full size:

   (a) A sectional FE in the direction of arrow B. The section should be parallel to the sides of the rod and pass through the centre of the hinge bolt.
   (b) A plan in the direction of arrow A and projected from the elevation. Show all hidden detail in this view: (i) the smallest diameter spigot on the bolt should be shown threaded M10 for 15 mm and the bolt should be fastened by an M10 nut; (ii) six main dimensions should be added to the views; (iii) print the title 'Safety valve operating link detail' in the bottom right-hand corner of your sheet in 6 mm letters; (iv) in the bottom left-hand corner of the sheet print the type of projection you have used.
   East Anglian Examinations Board

2. Figure 11.8 shows the details of the parts of a machine vice. The movable jaw rests on the bed of the vice. The thumbscrew is screwed into the body of the vice and enters the 8 mm hole in the movable jaw.

   The two pegs are fitted into the two holes in the movable jaw and secure in position the thumbscrew by the 5 mm diameter neck.
   (a) Draw, twice full size, the assembled vice with the jaws 12 mm apart and the Tommy bar of the thumbscrew in the vertical position: (1) an

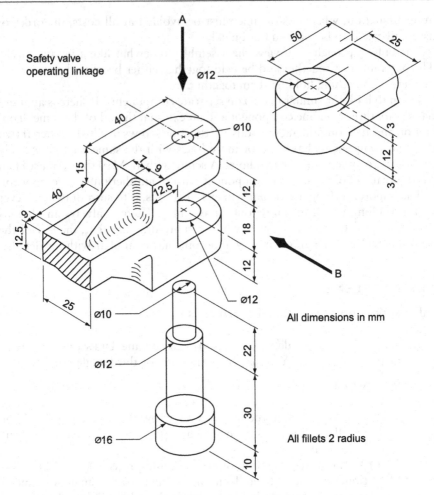

Safety valve operating linkage

All dimensions in mm

All fillets 2 radius

Figure 11.7 Detail from a stationary engine.

FE looking in the direction of arrow A; (2) a plan; (3) an elevation looking in the direction of arrow B.

First or third angle projection may be used.

Hidden details need not be shown and only five dimensions need be shown. (b) Make a good quality *freehand* sketch of the assembled vice of approximately twice full size.

Add the title, the scale and the angle of projection used, in letters of a suitable size.

Southern Regional Examinations Board

3 Parts of a step-down pulley are shown in Figure 11.9. Draw, full size, the following views of the step-down pulley fully assembled: (a) a sectional FE on the plane XX; (b) an EE when viewed in the direction of arrow A;

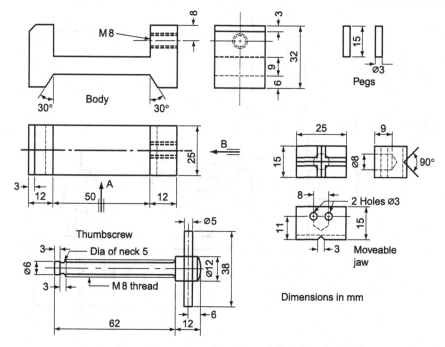

Figure 11.8 Parts of a machine vice.

(c) a plan projected from (a) above and looking in the direction of arrow D. Print the title. Size of letter to be 8 mm.

Print in the angle of projection you have used. Size of letter to be 6 mm. Print in the scale. Size of letter to be 4 mm.

No dimensions are required on your drawing: (1) radii at B 12 mm; (2) radii of fillets 6 mm; (3) use your own judgement to determine the size of any dimensions not given on the drawing; (4) no hidden details are to be shown.

Associated Lancashire Schools Examining Board

4 Figure 11.10 shows parts of a magnifying glass as used by engravers and biologists. The stem A screws to the base B. The stand and the glass C are connected by two link bars D which are held in the desired position by the distance piece F and two 6 mm round-headed bolts and wing nuts E.

Draw, full size, the following views of the magnifying glass and stand fully assembled: (a) an elevation in the same position as the encircled detail; (b) a plan projected from this elevation; (c) an elevation as seen in the direction of the arrow.

The link bars are inclined at 15° and the glass at 30° to the horizontal.

Include on this drawing eight of the main dimensions and a title block in the bottom right-hand corner. Within this block letter the title,

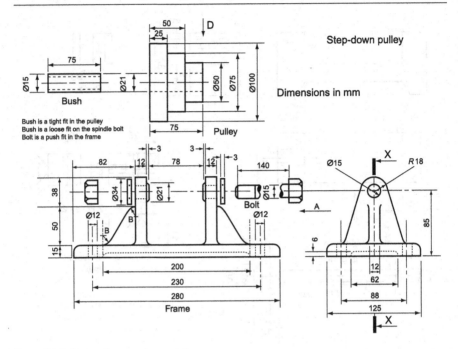

Figure 11.9 Parts of a step-down pulley.

MAGNIFYING GLASS AND STAND, and state the scale and your school and name.

Southern Universities' Joint Board

5 Figure 11.11 is an exploded view of a plummer block bearing. Draw, to a scale of 2:1 in first angle orthographic projection, the following views of the assembled bearing: (a) a sectioned elevation as seen looking in the direction of arrow X; the cutting plane to be vertical and to pass through AA; (b) an elevation as seen when looking from the left of view (a); (c) a plan (beneath view (a)).

The cap is held in position on the studs by means of single chamfered hexagonal nuts that have 2 mm thick single bevelled plain washers beneath them.

Although the tapped hole, for the lubricator, is shown in the cap, further details of this hole are not shown and have been left to your own discretion. Details of this hole are required to be shown, as are the details of the stud holes in the base.

Hidden detail of the base, *only*, is to be shown in view (c); no other hidden detail is to be shown.

Fully dimension *only* the bottom brass.

Draw a suitable frame around your drawing and insert the title PLUMMER BLOCK, your name, the scale and the system of projection.

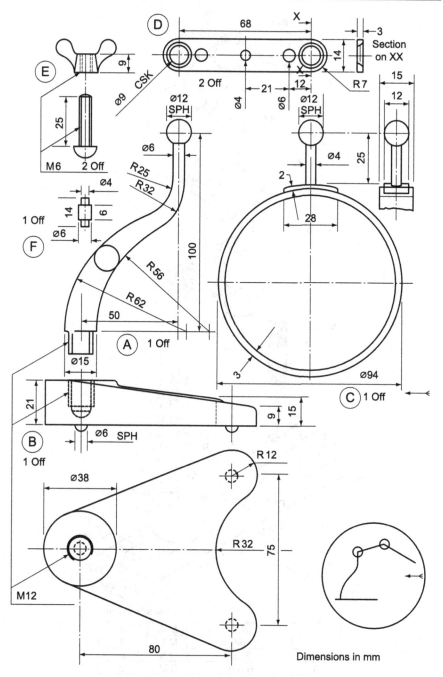

Figure 11.10 Parts of a magnifying glass.

Dimensions in mm

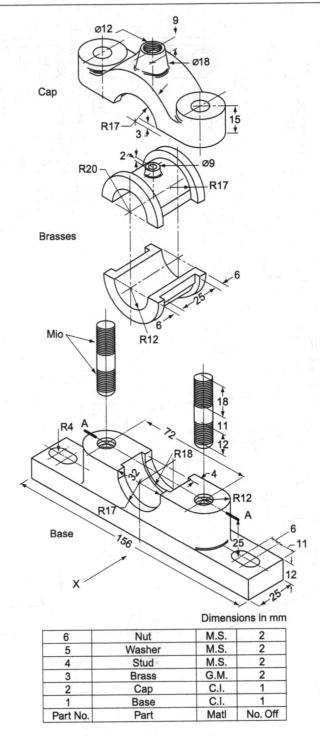

6	Nut	M.S.	2
5	Washer	M.S.	2
4	Stud	M.S.	2
3	Brass	G.M.	2
2	Cap	C.I.	1
1	Base	C.I.	1
Part No.	Part	Matl	No. Off

Figure 11.11 Exploded view of a plummer block bearing.

A parts list incorporating the part number, name of part and the quantity of each part is to be completed in the lower right-hand corner of your sheet of drawing paper immediately above your title block.
Oxford Local Examinations

6 Figure 11.12 shows the details of a small machine vice and the key to its assembly. Draw, full size and in correct orthographic projection, the following views of the completely assembled vice, the sliding jaw being approximately 25 mm from the fixed jaw. (a) A sectional elevation on a VP passing through the axis of the square-headed screw, in the direction indicated by XX in the key; (b) a plan projected from the sectional elevation (a).

Details of screws are not given, and these may be omitted from your drawing. No hidden edges are to be shown and dimensions are not required.
Either first or third angle (but not both) methods of projection may be used; the method chosen must be stated on the drawing.
In the bottom right-hand corner of the paper draw a title block 112 mm 62 mm and in it print neatly the drawing title, MACHINE VICE ASSEMBLY, the scale and your name.
University of London School Examinations

7 Figure 11.13 consists of a half-sectioned FE, a side elevation and a plan of part of a lathe steady. Draw, full size, in first angle orthographic projection, detail drawings of each of the six parts of the lathe steady, as follows:

Part 1. Top. An FE and an elevation as seen from the right of the FE.
Part 2. Base. An FE, a plan and a half-sectioned side elevation as seen from the right of the FE; the section planes for this view are PQ, QR.
Part 3. Pivot screw. An elevation with the axis horizontal and the head to the right.
Part 4. Hinge pin. An elevation with the axis horizontal and an EE as seen from the left.
Part 5. Eye bolt. An FE and an elevation as seen from the left of the FE.
Part 6. Clamping nut. A plan and elevation.
Full hidden detail is to be shown in Part 1 only; no other hidden detail is to be shown.
Part 5 is to be fully dimensioned; no other dimensions are to be shown.
Draw a suitable frame around your drawing and insert the title LATHE STEADY, your name, the scale and the system of projection.
Also insert, in the right-hand corner of the frame, above the title block, a parts list giving the part number, name of part and the quantity. It is suggested that the parts list should have a width of 125 mm.
Figure 11.14 gives an indication of how your sheet should be arranged.
Oxford Local Examinations

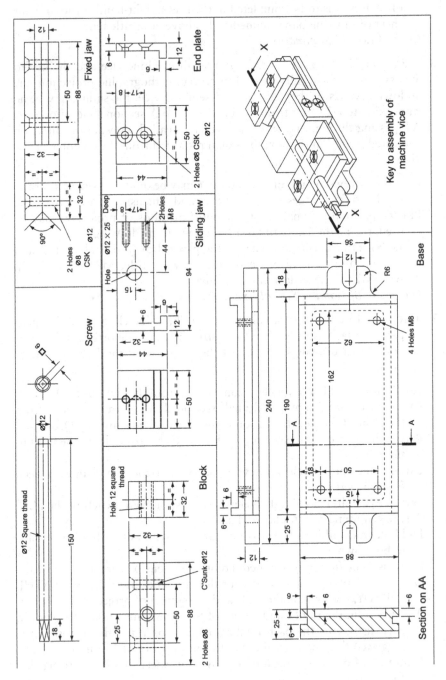

Figure 11.12 Details of a small machine vice.

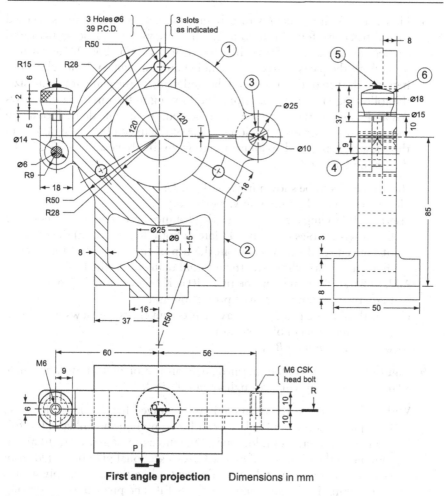

3 Holes Ø6
39 P.C.D.

3 slots
as indicated

R50

R15
R28

①

③

Ø25

Ø10

R15
R28

⑤

⑥

8

Ø18
Ø15

20

37

④

85

Ø14
Ø6
R9

18

R50
R28

Ø25
Ø9
15

②

8

16
37

R50

50

3

8

First angle projection   Dimensions in mm

60

56

M6
9

M6 CSK
head bolt

R

6

10
10

P

Figure 11.13 Lathe steady.

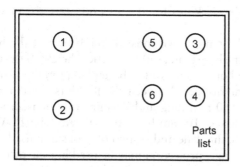

①

⑤   ③

②

⑥   ④

Parts
list

Figure 11.14 Arrangement of views to solve previous problem.

8   Figure 11.15 shows details of a bogie truck. The fixed plate is secured
    to the truck by four M10 countersunk screws (not to be shown) and
    carries the wheel and frame by means of a central bolt M18, 50 mm
    long. With the parts correctly assembled and allowing a 2 mm clearance
    between the top of the frame and the underside of the plate draw, full size,
    the following views: (a) a sectional FE looking in the direction of arrows
    AA and taken on the centre line as indicated; (b) an outside EE looking in
    the direction of arrow B, the longer side of the fixed plate to be shown in
    this view; (c) an outside plan projected from (b) above and looking in the
    direction of arrow C.

Hidden details to be shown in view (b) *only*.
The castle nuts to be shown in view (b) *only*.
Insert the following dimensions: (i) the distance between the centres of the
    fixing screw holes in the fixed plate in both directions; (ii) the distance
    of the centre of the wheel from the top of the frame; (iii) the outside
    diameter of the wheel; (iv) the internal diameter of the bush.
Add, in letters 10 mm high, the title TRUCK BOGIE, and in letters 6 mm
    high, the scale and system of projection used.
First or third angle projection may be used but the three views must be in
    a consistent system of projection.
Associated Examining Board

9   Figure 11.16 shows details in third angle projection of an adjustable angle
    plate capable of moving through angles 0°–90°.

With the parts correctly assembled and arranged for an angle of 0°, i.e.
    base plate and movable plate both horizontal, draw, full size, the fol-
    lowing views: (a) an FE looking in the direction of arrow A; (b) an EE
    looking in the direction of arrow C; (c) a sectional plan projected from
    (a), the section being taken on the centre line through the pivot pins
    and looking in the direction of arrows BB. The pivot pins are driving
    fits in the base plate bosses and sliding fits in the movable plate bosses.
    The pins are to be assembled with their 32 mm diameter heads placed
    on the 34 mm counterbore sides of the bosses. Hidden detail is to be
    shown in view (a) only.
Only the scale markings for 0°, 30°, 60° and 90° in view (a) are to be
    shown.
Insert the following dimensions: (i) the length of the base plate; (ii) the
    width of the base plate; (iii) the vertical height of the centre of the pivot
    above the bottom face of the base plate; (iv) the diameter of the pivot
    pin; (v) the distance between the tee slots in the movable plate. In a
    rectangle 150 mm long and 75 mm wide in a corner of the drawing
    insert, in letters 10 mm high, the title ANGLE PLATE and, in letters
    6 mm high, the scale and system of projection used.
First or third angle projection may be used but the three views must be in
    a consistent system of projection.
Associated Examining Board

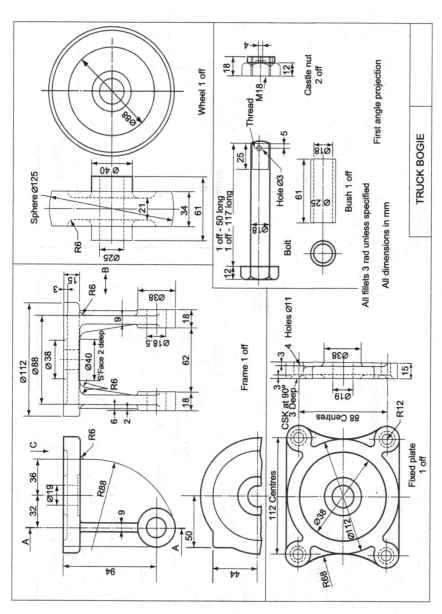

Figure 11.15 A bogie truck.

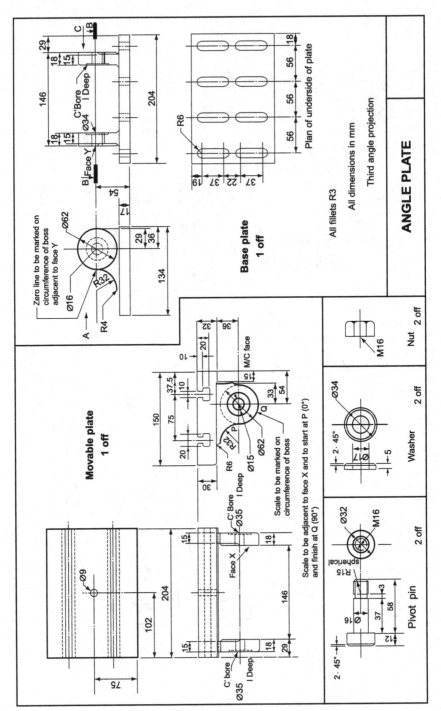

Figure 11.16 Adjustable angle plate.

Chapter 12

Some More Problems Solved by Drawing

This chapter introduces the student to some more drawing techniques. It should be emphasised that the topics are only introduced; all of them can be studied in much greater depth, and any solutions offered in this chapter will apply to simple problems only.

12.1 AREAS OF IRREGULAR SHAPES

It is possible to find, by drawing, the area of an irregular shape. The technique does not give an exact answer but, carefully used, can provide a reasonable answer. Look at Figure 12.1. The shape is trapezoidal with irregular end lines. A centre line has been drawn and you can see that the shaded triangles at the top and bottom are, in their pairs, approximately the same in area. Thus, the approximate area of the whole figure is the width W multiplied by the height at the centre. That height is called the 'mid-ordinate' and the whole technique is called the 'mid-ordinate rule'.

Figure 12.2 shows how it is applied to a larger figure.

The figure is divided into several equal strips (width W), in this case eight. The more strips that are drawn (within reason) the greater the accuracy of the final calculation. The centre line of each strip (the mid-ordinate) is drawn. Each of these lengths is measured and the area of the figure is given by

$$W\left(O_1 + O_2 + O_3 + O_4 + O_5 + O_6 + O_7 + O_8\right)$$

A sample of this technique is shown in Figure 12.3. The curve shown is a sine curve, the curve that emerges if you plot the values of the sine of all the angles from 0° to 90°.

The area is the product of the width of each strip and the sum of all the mid-ordinates.

12.2 RESOLUTION OF FORCES

All machinery, however simple, has forces acting on its parts. Buildings have forces acting on them: forces produced by the weight of the building itself, the weight of the things inside it and the wind pushing against it. An understanding

DOI: 10.1201/9781003001386-12

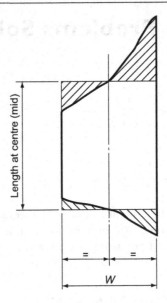

Figure 12.1 Irregular shape.

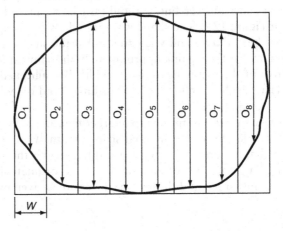

Figure 12.2 Irregular shape of a larger object.

of how these forces act and how they affect design is essential to a good draftsperson.

You must first understand the difference between stable and unstable forces and then study the effects of the unstable ones. The two men indulging in arm wrestling in Figure 12.4 are applying force. As long as the forces are equal, they will remain in the position shown. When one begins to apply more force than the other, the forces become unstable and the other has his hand forced back onto the table.

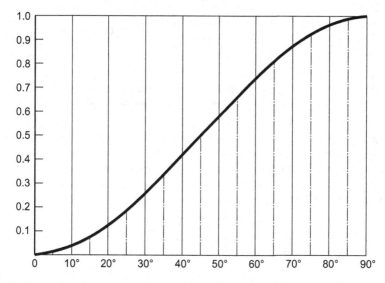

Figure 12.3 Sine curve.

Figure 12.4 Arm wrestling.

Some forces acting on a point are shown in Figure 12.5. The forces on the left are in line, they are equal and opposite and so the forces are stable. The centre forces are in line and opposite but one is larger than the other. Therefore the larger force will push the smaller force back. The forces on the right are

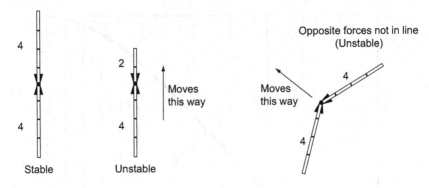

Figure 12.5 Forces acting on a point.

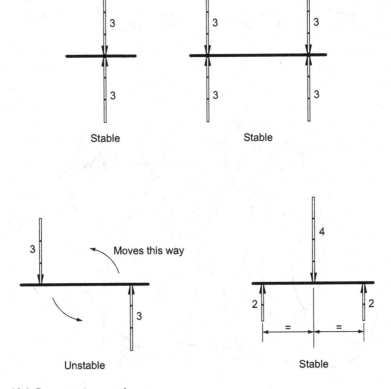

Figure 12.6 Forces acting on a beam.

equal in size but are not in line. The effect will be for the point to move in the direction shown.

Some forces acting on a beam are shown in Figure 12.6. The forces on the left are in line and opposite and so the forces are stable. The four forces on the second beam are also equal and opposite and so, although they are acting in pairs, the whole set of forces is stable. The forces acting on the third example are equal but not opposite. The effect of these forces acting in the way they are placed will be to rotate the beam as shown. This kind of force is important and is called a 'couple'. The final set of forces on the right are also stable. The two forces under the beam add up to the same total as the force above the beam. They are spaced at equal distance from the force above the beam and thus the whole system is stable.

Calculating Some of the Forces Acting on a Beam

If a beam is loaded, then, to stabilise it, forces have to be applied in opposition. These opposite forces are called 'reactions'. Some simple examples are shown in Figure 12.7. The loading is applied at a point and is therefore called a 'point load'.

The left-hand example has the reactions positioned at equal distances from the load and they will therefore be equal. In that case they must each be half the load.

The centre example has a load of 8 units[1] and one reaction is 5 units. For the loading and reaction system to be stable, the sum of the reactions must equal the loading. Therefore the reaction on the right (R_R) must be 3 units.

The left and right reactions are given in the example on the right. Since the sum of the reactions must equal the load, the load is 5 units.

Figure 12.8 shows two examples where the loads and reactions are known but the position of one reaction has to be calculated.

If the loaded system is stable then *all* the forces must balance. The reactions added together (8 + 4) must equal the load (12). The couples must also be balanced out. The left-hand couple, reaction times distance to the load (8 × 2), must equal the right-hand couple (4 × x). Thus the distance can be worked out in the simple equation shown.

The example on the right is similar.

The force called a couple in this chapter is also called a *moment* or *bending moment* (because it tries to bend the beam) by engineers.

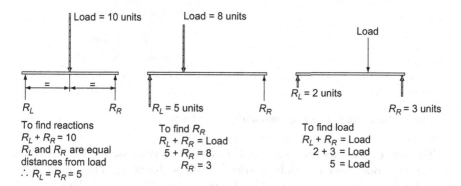

Figure 12.7 Reactions in a point loaded beam.

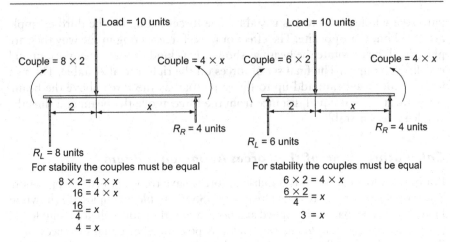

Figure 12.8 Finding the position of the reaction in a stable loaded beam.

Left diagram:

Load = 10 units

Couple = 8 × 2   Couple = 4 × x

2   x

R_R = 4 units

R_L = 8 units

For stability the couples must be equal

$$8 \times 2 = 4 \times x$$
$$16 = 4 \times x$$
$$\frac{16}{4} = x$$
$$4 = x$$

Right diagram:

Load = 10 units

Couple = 6 × 2   Couple = 4 × x

2   x

R_R = 4 units

R_L = 6 units

For stability the couples must be equal

$$6 \times 2 = 4 \times x$$
$$\frac{6 \times 2}{4} = x$$
$$3 = x$$

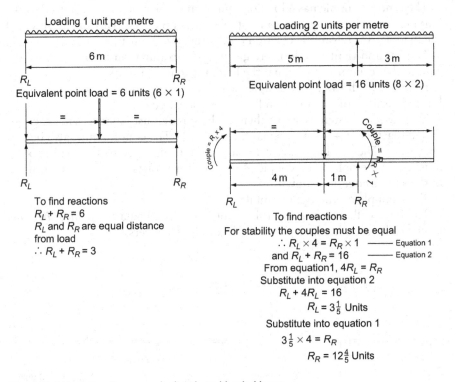

Figure 12.9 Reactions in an evenly distributed loaded beam.

Left diagram:

Loading 1 unit per metre

6 m

R_L R_R

Equivalent point load = 6 units (6 × 1)

R_L R_R

To find reactions
$R_L + R_R = 6$
R_L and R_R are equal distance from load
∴ $R_L + R_R = 3$

Right diagram:

Loading 2 units per metre

5 m   3 m

Equivalent point load = 16 units (8 × 2)

Couple = $R_L \times 4$ Couple = $R_R \times 1$

4 m   1 m

R_L R_R

To find reactions
For stability the couples must be equal
∴ $R_L \times 4 = R_R \times 1$ ———— Equation 1
and $R_L + R_R = 16$ ———— Equation 2
From equation 1, $4R_L = R_R$
Substitute into equation 2
$$R_L + 4R_L = 16$$
$$R_L = 3\tfrac{1}{5} \text{ Units}$$
Substitute into equation 1
$$3\tfrac{1}{5} \times 4 = R_R$$
$$R_R = 12\tfrac{4}{5} \text{ Units}$$

Two more examples are shown in Figure 12.9. In this case the loading is not acting at a point but is evenly distributed along the beam. This could make calculations difficult; in fact, we are able to change the loading. The total load is the load per metre multiplied by its length (6 × 1 = 6 units). The effect of this evenly distributed load on the beam is the same as if it were a point load acting at the centre. The lower left drawing in Figure 12.9 shows how this looks. The reactions can now be easily calculated.

The example on the right also has the evenly distributed load changed into a point load acting at the centre of the beam. The size of the two reactions can be calculated with a simple simultaneous equation (shown underneath the diagram).

Forces Acting at a Point

When two or more forces act at or on a point, it is useful to be able to change these forces and show how they could be replaced with a single force that acts in the same way. This force is called the 'resultant force'. The force that would have to be applied to stabilise the system (by acting against and cancelling out the two or more forces) is called the 'equilibrant force'. An example is shown in Figure 12.10.

The two forces are of 6 and 4 units. The resultant and equilibrant forces can be found by drawing lines parallel to these two forces and forming a parallelogram. The resultant and equilibrant forces are equal to the length of the longer diagonal of this parallelogram. An alternative is to draw triangles as shown in the figure. The result of these drawings is shown on the extreme right. The resultant force is the one that would have the same effects as the two forces if it replaced them; the equilibrant force acts against the two forces and makes the system stable.

Figure 12.11 shows how to find the equilibrant force to three forces acting at a point. Draw a line parallel to force 1 and equal (in scale) to its force. From the end of this line draw a line parallel to force 2 and equal (in scale) to its force. From the end of this line draw a line parallel to force 3 and equal (in scale) to its force. The line that closes the quadrilateral is equal to the size of the equilibrant force (to scale) and acts in the direction of the equilibrant force. It can be transferred back to the original drawing of the forces and stabilise the system.

Figure 12.12 shows how to find the equilibrant force to four forces acting at a point. Once again, the forces are drawn to scale parallel to the way they are acting at the point. The forces are each taken in turn and are considered

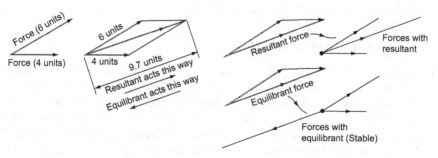

Figure 12.10 Forces acting at a point.

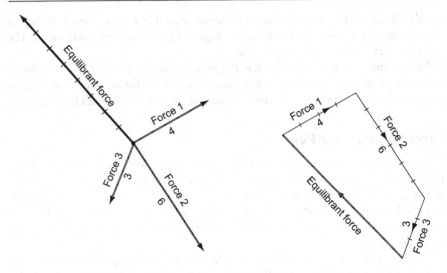

Figure 12.11 Finding the equilibrant force to three forces meeting at a point.

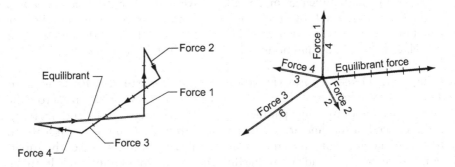

Figure 12.12 Finding the equilibrant force to four forces meeting at a point.

clockwise. The point to note in this example is that, although the lines cross each other, the equilibrant force is still the one that closes the figure, the one that joins the end of the line representing the last force drawn to the beginning of the line representing the first force drawn.

12.3 SIMPLE CAM DESIGN

Cams are used in machines to provide a controlled up and down movement. This movement is transmitted by means of a *follower*. A cam with a follower is shown in Figure 12.13 in the maximum 'up' and 'down' positions. The difference between these two positions gives the *lift* of the cam. The shape of the cam, its *profile*, determines how the follower moves through its lift and *fall*.

A cam is designed to make a pan of a machine move in a particular way. For example, cams are used to open and close the valves that control the petrol

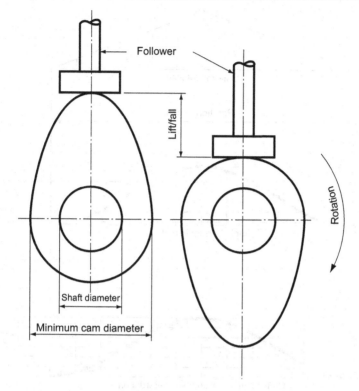

Figure 12.13 Simple cam and follower.

mixture going into and the exhaust gases coming out of an internal combustion engine. Obviously, the valves must open at the right time and at the right speed, and it is the cam that determines this. This control on the valve is exercised by the profile of the cam, determining how the follower lifts and falls. Three examples of types of lift and fall are shown in Figure 12.14.

The top diagram shows uniform velocity. This is a graph of a point that is moving at the same velocity 'upward' to halfway and then at the same velocity 'downward' to its starting point.

The centre diagram shows simple harmonic motion. This is the motion of a pendulum, starting with zero velocity, accelerating to a maximum and then decelerating to zero again (at the top of the curve) and then repeating the process back to the starting point. The curve is a sine curve and is plotted as shown.

The lower diagram shows uniform acceleration and retardation. In this case the acceleration is uniform to a point halfway up the lift and then retarding uniformly to the maximum lift. The process is then repeated in reverse back to the starting point. The curve is plotted as shown.

If a cam has to be designed, it is drawn around a specification. This must state the dimensions, the lift/fall and the *performance*. The performance states how the follower is to behave throughout one rotation of the cam. The designer

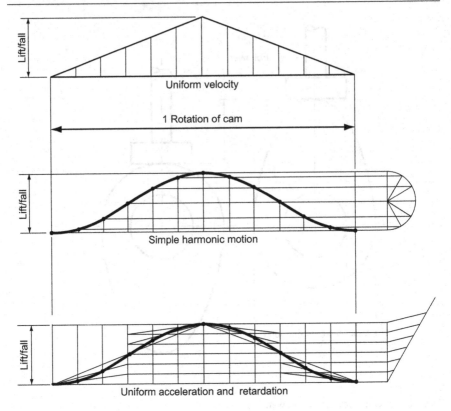

Uniform velocity

1 Rotation of cam

Simple harmonic motion

Uniform acceleration and retardation

Figure 12.14 Types of lift and fall.

must first draw the performance curve to the given specification. An example is shown in Figure 12.15. *Dwell* is a period when the follower is neither lifting nor falling.

The performance curve is started by drawing a base line of 12 equal parts, the total representing one rotation of the cam. The lift/fall is then marked out, and the performance curve drawn. The base line of the performance curve is then projected across and, with the centre line of the cam, forms the top of the circle representing the minimum cam diameter. Once the centre of the cam has been found, centre lines can be drawn at 30° intervals. The cam profile is plotted on these lines. Twelve points on the performance curve are then projected across to the centre line of the cam and then swung round with compasses to the intersecting points on the lines drawn at 30° intervals. If the cam is rotating clockwise the points 1 to 12 are marked out clockwise; if the rotation is anti-clockwise, as in this case, the points are marked out anti-clockwise.

Figure 12.16 shows another example of a cam design.

This is a more complicated profile than the previous example, but the method used to construct the profile is the same. This cam rotates in a clockwise direction.

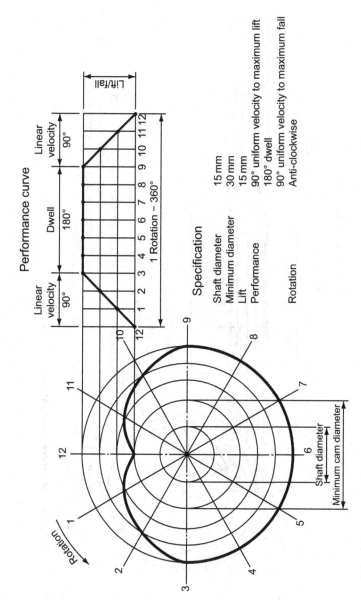

Figure 12.15 Example of cam design.

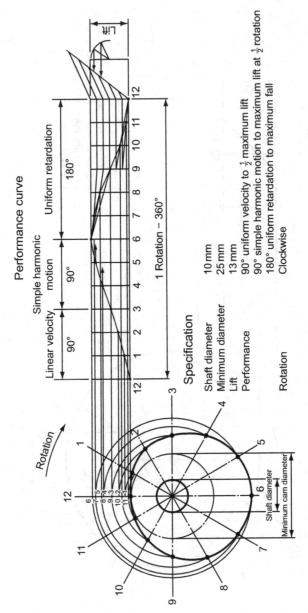

Figure 12.16 Example of a more complicated cam design.

Knife   Roller   Flat

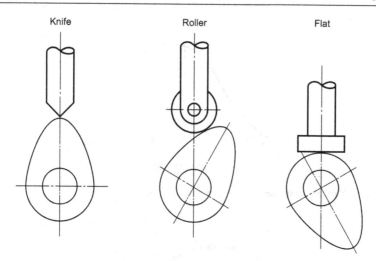

Figure 12.17 Three types of cam followers.

Finally, three different types of follower are shown in Figure 12.17. The left-hand example is a knife follower. It can be used with a cam that has a part of its profile convex, but it wears quickly. The centre example is a roller follower; this type of follower obviously reduces friction between the follower and the cam, something of a problem with the other types shown. The flat follower is an all-purpose type which is widely used. It wears more slowly than a knife follower does.

12.4 PROBLEMS

(All questions originally set in imperial units.)

1 The velocity of a vehicle moving in a straight line from start to rest was recorded at intervals of 1 minute, and these readings are shown in the table. Part of the velocity/time diagram is given in Figure 12.18.

Draw the complete diagram with a horizontal scale of 10 mm to 1 minute and vertical scale of 10 mm to 1 m/s. Using the mid-ordinate rule, determine the average velocity of the vehicle and hence the total distance travelled in the 12 minutes.

Time, t (min)	0	1	2	3	4	5	6	7	8	9	10	11	12
Velocity, v (m/s)	0	1.4	5.4	9.2	8.6	7.9	9.0	10.2	11.5	12.6	9.4	5.2	0

Oxford Local Examinations

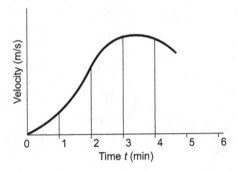

Figure 12.18 Velocity/time diagram for a moving vehicle.

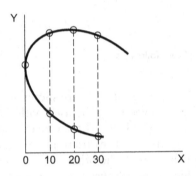

Figure 12.19 Indicator diagram from an engine test.

2   Figure 12.19 shows part of an indicator diagram which was made during an engine test. Draw the complete diagram, full size, using the information given in the table and then, by means of the mid-ordinate rule, determine the area of the diagram. Also, given that the area of the diagram can be given by the product of its length and average height, determine the average height and the average pressure if the ordinates represent the pressure to a scale of 1 mm to 60 kN/m^2.

OX(mm)	0	10	20	30	40	50
OY max (mm)	54	79	82	78	66	54
OY min (mm)	54	21	14	12	12	12
OX (mm)	60	70	80	90	100	
OY max (mm)	45	39	33	29	21	
OY min (mm)	12	12	12	13	21	

Oxford Local Examinations

3   Figure 12.20 shows four simply loaded beams. Find the size of the reactions marked X.

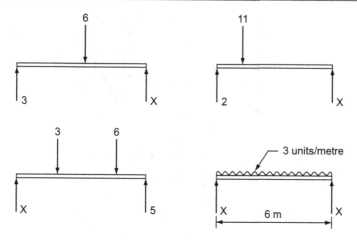

Figure 12.20 Loaded beams.

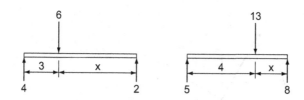

Figure 12.21 Loaded beams and their reactions.

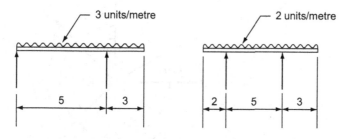

Figure 12.22 Loaded beams with evenly distributed loads.

4   Figure 12.21 shows two simply loaded beams with their reactions. Find the dimensions marked x for the loading to be in equilibrium (the couples to be equal).

5   Figure 12.22 shows two beams loaded with evenly distributed loads. The positions of the reactions are shown. Find the size of the reactions.

6   Figure 12.23 shows three examples of two forces acting at a point. Find the size of the resultant force for each example and measure the angular direction of the resultant force from the datum shown.

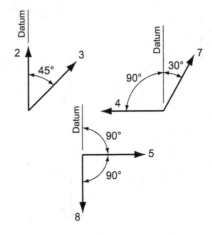

Figure 12.23 Examples of two forces acting at a point.

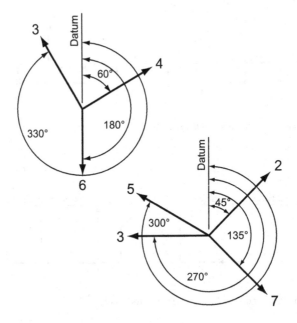

Figure 12.24 Examples of multiple forces acting at a point.

7   Figure 12.24 shows two examples of forces acting at a point. Find the size of the resultant force for both examples and measure and state the angular direction of the resultant force from the datum shown. From the same datum state the angular direction of the equilibrant force for both examples.

8 Plot the cam profile which meets the following specification:

Shaft diameter	15 mm
Minimum diameter	25 mm
Lift	12 mm
Performance	90° uniform velocity to maximum lift
	90° dwell
	180° uniform retardation to maximum fall
Rotation	Clockwise

Your cam profile must be drawn, twice full size.

9 Plot the cam profile which meets the following specifications:

Shaft diameter	12.5 mm
Minimum diameter	30 mm
Lift	12.5 mm
Performance	60° dwell
	90° simple harmonic motion to half lift
	30° dwell
	60° uniform acceleration to maximum lift
	120° uniform velocity to maximum fall
Rotation	Anti-clockwise

Your cam profile is to be drawn, twice full size.

NOTE

1 Forces in the SI system are given in newtons (the force required to accelerate 1 kilogram at 1 metre per second^2). For simplicity the loads are called units in this chapter.

Appendix A: ISO Standards Mentioned in This Book

ISO 3	*Preferred numbers – Series of preferred numbers*
ISO 14	*Straight-sided splines for cylindrical shafts with internal centering – Dimensions, tolerances and verification*
ISO 68–1	*ISO general purpose screw threads – Basic profile – Part 1: Metric screw threads*
ISO 128–1	*Technical product documentation (TPD) – General principles of representation – Part 1: Introduction and fundamental requirements*
ISO 128–2	*Technical product documentation (TPD) – General principles of presentation – Part 2: Basic conventions for lines*
ISO 128–3	*Technical product documentation (TPD) – General principles of presentation – Part 3: Basic conventions for views*
ISO 129–1	*Technical product documentation (TPD) – Presentation of dimensions and tolerances*
ISO 216	*Writing paper and certain classes of printed matter – Trimmed sizes – A and B series, and indication of machine direction*
ISO 273	*Fasteners – Clearance holes for bolts and screws*
ISO 286–1	*Geometrical product specifications (GPS) – ISO code system for tolerances on linear sizes – Part 1: Basics of tolerances, deviations and fits*
ISO 286–2	*Geometrical product specifications (GPS) – ISO code system for tolerances on linear sizes – Part 2: Tables of standard tolerance classes and limit deviations for holes and shafts*
ISO 866	*Centre drills for centre holes without protecting chamfers – Type A*
ISO 1101	*Geometrical product specifications (GPS) – Geometrical tolerancing – Tolerances of form, orientation, location and run-out*
ISO 1302	*Geometrical product specifications (GPS) – Indication of surface texture in technical product documentation*
ISO 2162–1	*Technical product documentation – Springs – Part 1: Simplified representation*
ISO 2203	*Technical drawings – Conventional representation of gears*
ISO 2540	*Centre drills for centre holes with protecting chamfer – Type B*

ISO 2541   *Centre drills for centre holes with radius form – Type R*
ISO 2768–1   *General tolerances – Part 1: Tolerances for linear and angular dimensions without individual tolerance indications*
ISO 3040   *Geometrical product specifications (GPS) – Dimensioning and tolerancing – Cones*
ISO 3098   *Technical product documentation – Lettering*
ISO 4156–2   *Straight cylindrical involute splines – Metric module, side fit – Part 2: Dimensions*
ISO 4287   *Geometrical product specifications (GPS) – Surface texture: Profile method – Terms, definitions and surface texture parameters*
ISO 4288   *Geometrical product specifications (GPS) – Surface texture: Profile method – Rules and procedures for the assessment of surface texture*
ISO 5455   *Technical drawings – Scales*
ISO 5456–2   *Technical drawings – Projection methods – Part 2: Orthographic representations*
ISO 5456–3   *Technical drawings – Projection methods – Part 3: Axonometric representations*
ISO 5456–4   *Technical drawings – Projection methods – Part 4: Central projection*
ISO 5457   *Technical product documentation – Sizes and layout of drawing sheets*
ISO 5459   *Geometrical product specifications (GPS) – Geometrical tolerancing – Datums and datum systems*
ISO 6284   *Construction drawings – Indication of limit deviations*
ISO 6410–1   *Technical drawings – Screw threads and threaded parts – Part 1: General conventions*
ISO 6410–3   *Technical drawings – Screw threads and threaded parts – Part 3: Simplified representation*
ISO 6411   *Technical drawings – Simplified representation of centre holes*
ISO 6413   *Technical product documentation – Representation of splines and serrations*
ISO 7083   *Technical product documentation – Symbols used in technical product documentation – Proportions and dimensions*
ISO 7200   *Technical product documentation – Data fields in title blocks and document headers*
ISO 7573   *Technical product documentation – Parts lists*
ISO 8826–1   *Technical drawings – Rolling bearings – Part 1: General simplified representation*
ISO 8826–2   *Technical drawings – Rolling bearings – Part 2: Detailed simplified representation*
ISO 9222–1   *Technical drawings – Seals for dynamic application – Part 1: General simplified representation*
ISO 9222–2   *Technical drawings – Seals for dynamic application – Part 2: Detailed simplified representation*

ISO 10209	*Technical product documentation – Vocabulary – Terms relating to technical drawings, product definition and related documentation*
ISO 13444	*Technical product documentation (TPD) – Dimensioning and indication of knurling*
ISO 13715	*Technical product documentation – Edges of undefined shape – Indication and dimensioning*
ISO 14405	*Geometrical product specifications (GPS) – Dimensional tolerancing*
ISO 15786	*Technical drawings – Simplified representation and dimensioning of holes*
ISO 29845	*Technical product documentation – Document types*
ISO 81714–1	*Design of graphical symbols for use in the technical documentation of products – Part 1: Basic rules*

Appendix B: Representation of Threaded Fasteners

Description	Simplified representation	Description	Simplified representation
Hexagon head screw		Countersunk screw, cross slotted	
Square head screw		Setscrew, slotted	
Hexagon socket screw		Wood and self-tapping screw, slotted	
Pan head screw, slotted		Wing screw	
Cheese head screw, cross slotted		Hexagon nut	
Raised countersunk screw, slotted		Slotted nut	
Raised countersunk screw, cross slotted		Square nut	
Countersunk screw, slotted		Wing nut	

Figure B.1 Simplified representation of threaded fasteners according to ISO 6410–3.

Index

Printed in the United States
by Baker & Taylor Publisher Services